ACARINE BIOCONTROL AGENTS

STRATHCLYDE UNIVERSITY LIBRARY

30125 00382240 9

This book is to be returned on or before
the last date stamped below.

LIBREX —

ACARINE BIOCONTROL AGENTS

An illustrated key and manual

URI GERSON
The Hebrew University of Jerusalem
Rehovot,
Israel

and

ROBERT L. SMILEY
US Department of Agriculture
Beltsville,
Maryland,
USA

CHAPMAN AND HALL
LONDON · NEW YORK · TOKYO · MELBOURNE · MADRAS

UK	Chapman and Hall, 11 New Fetter Lane, London EC4P 4EE
USA	Chapman and Hall, 29 West 35th Street, New York NY10001
JAPAN	Chapman and Hall Japan, Thomson Publishing Japan, Hirakawacho Nemoto Building, 7F, 1-7-11 Hirakawa-cho, Chiyoda-ku, Tokyo 102
AUSTRALIA	Chapman and Hall Australia, Thomas Nelson Australia, 480 La Trobe Street, PO Box 4725, Melbourne 3000
INDIA	Chapman and Hall India, R. Sheshadri, 32 Second Main Road, CIT East, Madras 600 035

First edition

© 1990 Uri Gerson and Robert L. Smiley

Typeset in 10/12 Palatino by
Photoprint, Torquay, Devon
Printed in Great Britain by
St Edmundsbury Press, Bury St Edmunds, Suffolk.

ISBN 0 412 36060 8

All rights reserved. No part of this publication may be reproduced or transmitted, in any form or by any means, electronic, mechanical, photocopying, recording or otherwise or stored in a retrieval system of any nature, without the written permission of the copyright holder and the publisher, application for which shall be made to the publisher.

British Library Cataloguing in Publication Data

Gerson, Uri
  Acarine biocontrol agents.
  1. Pests. Biological control
  I. Title II. Smiley, Robert L.
  628.9'6

ISBN 0-412-36060-8

Library of Congress Cataloging-in-Publication Data

Gerson, U.
  Acarine biocontrol agents : an illustrated key
  and manual / Uri Gerson and Robert L. Smiley.
    p.   cm.
  Includes bibliographical references.
  ISBN 0-412-36060-8
  1. Mites as biological pest control agents. 2. Mites-
  -Identification. I. Smiley, Robert L. II. Title.
  SB976.M58G47 1990
  632'.96—dc20   89-48407
                                                    CIP

Contents

Colour plates appear between pages 54 and 55
Acknowledgements vii

Foreword by Lloyd Knutson ix

Introduction 1
    1   History and current work 5

Part One   Illustrated Key and Family Descriptions 13

    2   Illustrated key to relevant acarine families 15
    3   Acaridae 46
    4   Anystidae 48
    5   Arrenuridae 50
    6   Ascidae 52
    7   Bdellidae 54
    8   Camerobiidae 56
    9   Cheyletidae 58
    10   Cunaxidae 61
    11   Eriophyidae 62
    12   Erythraeidae 64
    13   Eupalopsellidae 66
    14   Galumnidae 68
    15   Hemisarcoptidae 70
    16   Hydryphantidae 72
    17   Laelapidae 74
    18   Limnesiidae 76
    19   Macrochelidae 78
    20   Parasitidae 81
    21   Phytoseiidae 83
    22   Pionidae 86
    23   Podapolipidae 88
    24   Pterygosomatidae 91
    25   Pyemotidae 93

CONTENTS

26	Stigmaeidae	96
27	Tarsonemidae	98
28	Tetranychidae	100
29	Trombidiidae	102
30	Tydeidae	104
31	Uropodidae	106

Part Two Acari as Biocontrol Agents 109

32	Acari as natural enemies of nematodes	111
33	Mites which affect grasshopper and locust populations	112
34	Acari as natural enemies of scale insects	114
35	Acari as natural enemies of aquatic Diptera of medical importance	115
36	Acarine biocontrol agents in stored products	117
37	Influence of host plants on the efficacy of acarine biocontrol agents	119
38	The effect of pesticides on acarine biocontrol agents	123
39	Rearing and shipping	129
40	Demonstrating the efficacy of acarine biological agents	133
41	Attributes of efficient acarine biocontrol agents	141
42	References	148
43	Recommendations for further reading	162
	Index	167

Acknowledgements

This book could not have been written without the help willingly extended to us by many colleagues. We wish to extend our heartfelt gratitude to Lloyd Knutson, Director, Biological Control of Weeds Laboratory, Europe, United States Department of Agriculture (USDA) and Agricultural Research Service (ARS), Rome, for his encouragement upon initiating this study, for his many valuable suggestions, and for writing the Foreword; and to R.L. Luck, Department of Entomology, University of California, Riverside, for his gracious hospitality and for his many helpful suggestions. We also wish to thank the following for reviewing the manuscript and for offering useful comments: E.W. Baker, Collaborator, and members of the peer review committee, S. Nakahara and P.V. Peterson (all at the Systematic Entomology Laboratory, ARS, USDA, Beltsville, Maryland); J.R. Gorham, Research Entomologist, Food and Drug Administration, Washington, DC, and J.A. McMurtry, Department of Entomology, University of California, Riverside. Special thanks are due to D.R. Cook, formerly Biology Department, Wayne State University, Detroit, and to G.R. Mullen, Department of Zoology-Entomology, Auburn University, Auburn, for their advice and illustrations on water mites. We would also like to thank those who kindly made various photographs available to us: J. Moser, Forest Service, USDA, Pineville; R.F.W. Schroder, Beneficial Insects Laboratory, ARS, USDA, Beltsville; M.A. Hoy and J.K. Clark, Department of Entomological Sciences, University of California, Berkeley; and Y.S. Chow, Institute of Zoology, Academia Sinica, Taipei, Taiwan.

Foreword

There is urgent need to increase the use of biological methods of pest control throughout the world. Pesticide resistance and residues, continuing establishments of foreign pests in new areas, resurgence of native pests, the critically important ground water quality issue, environmental concerns such as the impact of pest management on endangered species, increasing pest problems associated with increasing acreage in minimum-tillage and land retirement programmes, and economics – all of these issues are major driving forces in the search for new biological control agents. Mites are among the major, and relatively unutilized, resources for biological control agents for practically all classes of pests.

Having been involved in the development of the 1982 Berkeley conference on biological control of pests by mites, it is gratifying to see, in the presentation of this work, another major step forward. Although there have been some outstandingly successful control programmes and critically important research studies on mites as biological control agents, the field has languished because of lack of organized information; shortage of trained personnel; insufficient communication between acarologists and biological control workers; lack of a predictive, theoretical framework; and grossly inadequate knowledge of the systematics of mites.

Uri Gerson and Robert Smiley have brought together and analysed a wealth of information in this book and have presented it in a manner that will be of great value to the biological control worker and the acarologists, the specialist and non-specialist, the field-person and the theoretician.

LLOYD KNUTSON

Director, Biological Control of Weeds Laboratory, Europe
US Department of Agriculture, Agricultural Research Service

Introduction

'Biological control provides an environmentally safe, cost-effective, and energy-efficient means of pest control, either alone or as a component of integrated pest management. The predatory mites in the family Phytoseiidae are examples of biological control agents that have been recognized only recently as effective components of agricultural systems. The relative slowness with which this fact was recognized suggested to us that other mite groups may be overlooked at present but be capable of serving as effective biological control agents as soon as we gain additional knowledge' (Hoy *et al.*, 1983). This opening paragraph of the introduction to the first-ever conference on the biological control of pests by mites is also the starting point of the present publication.

Hoy *et al.* (1983) noted that in the past little communication took place between acarologists and biological control practitioners; the conference had in fact been convened in order to serve as a meeting ground for these two groups of scientists. The wide choice of topics discussed attested to some advances made and to many goals yet to be attained. One of the major needs identified was the basic recognition of mites that could serve as biological control agents. Professional entomologists, extension officers and other personnel engaged in economic entomology (including medical and veterinary entomology) often record mites found associated with pests. However, many entomologists, untrained in acarology, cannot identify these mites, and thus they totally ignore them or, at best, refer them to overworked mite specialists. Answers are usually long in coming and possible interest raised in the observer may be dissipated by that time. The present volume is intended to serve as an introduction to the employment of mites in the biological control of pests. We begin with a brief historical review and discussion of some current work, and then present keys to all acarine families which include members known or believed to be of significance in pest control (including weed control). Each family is briefly characterized in the text in manual style and accompanied by a drawing, followed by discussion of the potential of its members in the biocontrol of pests. We round out the book with a few chapters which either deal with topics not included in Hoy *et al.* (1983), or else are intended to add to these earlier presentations.

Pests, in the present context, are invertebrates, such as insects, mites

INTRODUCTION

and nematodes, which harm their plant or animal hosts to some economic measure (including disease transmission), impair human health and/or are nuisances. Also included are weeds, plants which successfully repress, or compete with, agricultural crops, thus causing yield reductions and which may clog up waterways, unless appropriately controlled. Biological control of pests by mites, in the present context, means acarine activity (parasitism, predation, competition, plant feeding, disease transmission, other activities or any combination) which reduces pest populations and/or the extent of their damage to below accepted economic (or medical) injury levels. Such control comprises 'classical' biological control – introductions of natural enemies – as well as other strategies, such as augmentation and conservation. Furthermore, we include herein 'natural' as well as 'manipulated' biological control (brought about by human intervention).

This book does not include all mite families which have ever been recorded as being associated with pests (including weeds). Only those families that have one or more members which are known or postulated to have a detrimental effect on pest populations are included, making a total of 29 families. Members of many additional acarine families are known to prey on or parasitize pests (Smiley and Knutson, 1983), but data on their effect (if any) on prey/host populations are not available, and these families are not included. Cases in point are the soil-dwelling mesostigmatic Rhodacaridae and the water mite prostigmatic Sperchontidae. The former are known to feed on nematodes (Karg, 1983) and on various soil-inhabiting stored-product pest mites (Lee, 1974), but the only quantitative demonstrations of their feeding rates were within pot experiments (Sharma, 1971; van de Bund, 1972). A member of the Sperchontidae was believed by Davies (1959) to have some controlling effect on one group of black flies (Diptera: Simuliidae), a group whose members are pests of secondary nuisance importance, and Semushin (1981) reported that *Sperchon*-parasitized hungry black flies died slightly quicker than uninfested simuliids. Also excluded are the more common but apparently harmless phoretics, including the numerous ant, termite, and beetle associates, as well as many others. Nor did we include Acari which are natural enemies of non-pest insects and mites, and non-weed plants. Much of those data can be found in general reviews about insect–mite associations (Binns, 1982; Costa, 1969 (who cited some earlier literature); Hunter and Rosario, 1988; Karg and Mack, 1986; Lindquist, 1975; MacNulty, 1971; Samšiňak, 1966). Thompson and Simmonds (1965) catalogued the mites associated with pests. The first critical overview on the role of mites in the biological control of mite and insect pests was presented by Gerson and van de Vrie (1979), and Xin (1985) published a short review in Chinese. This topic was greatly elaborated

INTRODUCTION

upon at the aforementioned special conference, which dealt mostly with North American research (Hoy et al., 1983).

The 29 families discussed here probably represent only some of the mite families which can and do play a role in biological pest control. The discipline of biological control, as noted, is concerned with the manipulation of natural enemies of pests, and a necessary first step is the ability to recognize these beneficial organisms. In regard to the Acari, it is our feeling (substantiated by statements made by several participants in the aforementioned conference, Hoy et al., 1983), that most work is still at the initial, exploratory stage, i.e. recognizing the natural enemies (Gerson and van de Vrie, 1979). It is our hope that the present essay will promote further research concerned with exploring, developing and realizing the potential of mites in the biological control of pests.

REFERENCES

Binns, E.S. (1982) Phoresy as migration – some functional aspects of phoresy in mites. *Biol. Rev.*, **57**, 571–620.

Costa, M. (1969) The associations between mesostigmatic mites and coprid beetles. *Acarologia*, **11**, 411–28.

Davies, D.M. (1959) The parasitism of black flies (Diptera, Simuliidae) by larval water mites mainly of the genus *Sperchon*. *Can. J. Zool.*, **37**, 353–69.

Gerson, U. and van de Vrie, M. (1979) The potential of mites in the biological control of mite and insect pests. In *Proc. 4th Int. Congr. Acarol.* (ed. E. Piffl), pp. 629–35.

Hoy, M.A., Cunningham, G.L. and Knutson, L. (eds) (1983) *Biological Control of Pests by Mites*, University of California Special Publication no. 3304, 185 pp.

Hunter, P.E. and Rosario, R.M.T. (1988) Associations of mesostigmata with other organisms. *Ann. Rev. Entomol.*, **33**, 393–417.

Karg, W. (1983) Verbreitung und Bedeutung von Raubmilben der Cohors Gamasina als Antagonisten von Nematoden. *Pedobiologia*, **25**, 419–32.

Karg, W. and Mack, S. (1986) Bedeutung und Nutzung oligiphager Raubmilben der Cohors Gamasina Leach. *Arch. Phytopathol. Pflanzenschutz, Berlin*, **22**, 107–18.

Lee, D.C. (1974) Rhodocaridae (Acari: Mesostigmata) from near Adelaide, Australia. III. Behaviour and development. *Acarologia*, **16**, 21–44.

Lindquist, E.E. (1975) Associations between mites and other arthropods in forest floor habitats. *Can. Entomol.*, **107**, 425–37.

MacNulty, B.J. (1971) An introduction to the study of Acari–Insecta associations. *Proc. Trans. Br. Entomol. Soc.*, **4**, 46–70.

Samšiňak, K. (1966) Relations between mites and insects. *Zesz. Probl. Postepow nauk Rolniczych*, **65**, 77–87.

INTRODUCTION

Semushin, R.D. (1981) Water mites (Sperchontidae), parasites of black flies (Simuliidae). *Parazitologiia*, **15**, 27–30. (In Russian).

Sharma, R.D. (1971) Studies on the plant parasitic nematode *Tylenchorhynchus dubius*. *Meded. Landbouww. Wageningen*, **71**, 1–154.

Smiley, R.L. and Knutson, L. (1983) Aspects of taxonomic research and services relative to mites as biological control agents. In *Biological Control of Pests and Mites* (eds M.A. Hoy, G.L. Cunningham and L. Knutson), University of California Special Publication no. 3304, pp. 148–64.

Thompson, W.R. and Simmonds, F.J. (1965) *A Catalogue of the Parasites and Predators of Insect Pests*, Section 4, Host Predator Catalogue. Commonwealth Agricultural Bureaux.

van de Bund, C.F. (1972) Some observations on predatory action of mites on nematodes. *Zesz. Probl. Postepow nauk Rolniczych*, **129**, 103–10.

Xin, J. (1985) Current status and perspective in mites as biological control agents of insect pests. *Chin. J. Biol. Cont.*, **1**, 40–43.

1
History and current work

The first mite recognized for its ability to reduce pest populations appears to have been *Hemisarcoptes malus* (Shimer), feeding on the oystershell scale, *Lepidosaphes ulmi* (L.). Its discoverer (Shimer, 1868a) believed that '. . . it can be seen that, although it [*H. malus*] may not entirely exterminate the Apple Bark-louse [*L. ulmi*], yet it is exerting a wholesome, restraining influence, doing much more than man could do'. After another year's observations, Shimer (1868b) became more enthusiastic:

> . . . the young [mites] appear abundantly in the early spring, and destroy many young bark lice . . . This acarian is the most formidable enemy of the apple bark louse extant, and is far more effectual than all other combined restraints in nature, the most zealous efforts of man included . . . Last year when I discovered it, I considered it of great importance; but now after fully completing its natural history, I know that its importance is much greater than I then anticipated.

The importance of this natural enemy was reiterated by Riley (1873), and may well have been in the background of the first international shipment of natural enemies (which were mites) undertaken that year. The grape phylloxera (*Daktulosphaira vitifolii* Fitch) had recently become a major pest of grapes in France. Riley (1874a) who had just found an acarid mite ('*Tyroglyphus phylloxerae*' Riley and Planchon) feeding on this root-infesting pest in the USA, believed that the predator, 'renders efficient aid in keeping it [the grape phylloxera] in check in this country' (Riley, 1874b). Riley, therefore, shipped the mites to France, where they arrived safely, were released, became established, but failed to have any effect on the phylloxera (Howard, 1930). Riley later (1878) recognized several predaceous mites which fed on the eggs of pests. *Hemisarcoptes malus* was transferred in an early (1917) introduction effort from eastern to western Canada in order to control *L. ulmi* (Tothill, 1918). This project was rated as a successful biological control project by Turnbull and Chant (1961).

Pyemotes (=*Pediculoides*) was recognized in the 1880s as a natural enemy of many insect pests (reviewed by Webster, 1910). One species was mass-reared and tried against the cotton boll weevil (*Anthonomus grandis* Boheman) in Mexico (Rangel, 1901); subsequently it was introduced into Texas in 1902 to combat the same pest (Hunter and Hinds, 1904). The mites failed to penetrate cotton squares within which the pest feeds, and the effort was abandoned. The history of the use of *Pyemotes* against bark beetles was reviewed by Lindquist (1969).

The role of cheyletid mites in reducing populations of pestiferous acarids in stored products was apparently first recognized by Ewing (1912). This author reported that '. . . in a short period of only a few days this predaceous species [*Cheyletus*] had multiplied and destroyed about 95% of the pernicious Tyroglyphids' (for a review see Solomon, 1946). One species, *Cheyletus eruditus* (Schrank), was used in early theoretical work on predator–prey interactions (Gause *et al.*, 1936).

Weed control with Acari was first considered in 1924 when the spider mite *Tetranychus desertorum* Banks was found to have been accidentally introduced into Australia and to have become one of several feeders on a prickly-pear cactus (*Opuntia inermis* DC). The mite was initially believed to have some potential for biocontrol, as heavy infestations restricted cactus fruiting. However, it was later superseded by more efficient natural enemies (Hill and Stone, 1985). A further Australian discovery of an acarine biocontrol agent was the bdellid mite, *Bdellodes lapidaria* Kramer, recognized as a predator of the lucerne flea, the collembolan *Sminthurus viridis* (L.), a major pest of pastures (Womersley, 1933). This mite was subsequently introduced into South Africa and exerted control over another collembolan pest, namely *Bourletiella arvalis* Fitch (Wallace and Walters, 1974).

Howard *et al.* (1912) reviewed early observations on the parasitism of mosquitoes by water mites; the prevailing consensus at the time was that these mites were not especially harmful to their hosts. Uchida and Miyazaki (1935) appear to have been the first to consider water mites as important enemies of mosquitoes, based on observations that insects bearing more than four mites 'were inactive and did not attack human beings'.

'The economic value of mites as predatory agents in the control of bark beetles has never been fully realized' wrote Rust in 1933, and, despite much research on this topic (reviewed by Kinn, 1983), such 'realization' has not yet been achieved.

This brief review of early biocontrol efforts with mites serves to emphasize the fact that Acari have been considered and employed against a great variety of pests (including weeds) from the beginning of the rise of biological control as a discipline. Despite some success, however, it was only from the 1950s onwards that the potential of the Phytoseiidae and other mites was recognized and began to draw the attention which we feel they deserve as biocontrol agents. Three other important, on-going biocontrol projects that use mites and which began around that period remain to be noted. These include the search for acarine natural enemies of bark beetles (Kinn, 1983; Lindquist, 1969; Moser, 1975), the integration of predaceous mites into filth flies control programmes (Axtell, 1969; Axtell and Rutz, 1986), and the use of bdellids as predators of pestiferous collembolans (Wallace and Walters, 1974). One measure of the increasing volume of research devoted to mites as biocontrol agents might be gleaned from scanning the proceedings of the International Congresses of Acarology. In the first congress (Anonymous, 1964) there was no section on biological control, and only four papers pertaining to this topic were read. At the second congress (Evans, 1969) there was a separate biological control section, with five talks, and three additional papers were read at the Plant Mites section. The proceedings of the sixth congress (Griffiths and Bowman, 1984) included more than a dozen papers in the section on biological control and there were additional presentations elsewhere. It should be noted that in some of these sections no clear separation was made between biological control of mites and biological control by mites; students may thus be somewhat confused, like the narrator of Richard Hull's 'The Murder of my Aunt'.

Interest in the theoretical aspects of biological control by mites has stimulated some recent predator–prey studies, in the mode initiated by Gause *et al.* (1936). These studies, among others, include those of Burnett (1977), Kaiser (1983), Nachman (1987, and other papers), and Saito (1986, 1987). Dicke and Sabelis (1988) have developed an infochemical terminology based on interactions between predatory mites, their phytophagous mite prey and the host plants. Mite phoresy as migration was reviewed by Binns (1982), who recognized the implications of acarine dispersal for biological control.

Continuing research on acarine biocontrol agents centres around the Phytoseiidae. The demonstration of their ability to control spider mite pests has engendered a vast amount of scientific endeavour, including systematic tomes, life history and behaviour studies, the development of mass-rearing methods and efforts to select and release pesticide-resistant strains of promising species. Much of these data were summarized by Hoy (1982), Sabelis (1985) and McMurtry and Rodriguez (1987). A more recent development is the effort to try these mites against whiteflies (Homoptera: Aleyrodidae) (Meyerdirk and Coudriet, 1985) and thrips (Thysanoptera) (Hansen, 1988; Ramakers and van Lieburg, 1982; Tanigoshi *et al.*, 1984).

Current applied investigations include the introduction of macrochelid mites into Australia for filth fly control (Halliday and Holm, 1987), *Hemisarcoptes* into New Zealand against armoured scale insects on kiwifruit (M.G. Hill, unpublished), and phytoseiids into Africa for cassava green mite (*Mononychellus tanajoa* (Bondar)) control (Megevand *et al.*, 1987), mass-production and application of pyemotid mites for imported fire ant control (Thorvilson *et al.*, 1987), the augmented use of cheyletid mites for store-product acarid control (Žďárková, 1986), the search for mite parasites of the Colorado potato beetle (*Leptinotarsa decemlineata* (Say)) (Logan *et al.*, 1987) and the Mexican bean beetle (*Epilachna varivestis* Mulsant) (Hochmuth *et al.*, 1987), the discovery that tydeids may control the tomato russet mite, *Aculops lycopersici* (Massee) (Hessein and Perring, 1986) and that Anystidae are capable of killing tick larvae in the laboratory (Holm and Wallace, 1989). New, promising aspects of the use of mites for pest control include the emerging realization that various general acarine predators probably play an important role in the suppression of various pests in the soil (Brust and House, 1988; Walter *et al.*, 1988), that cheyletids may reduce populations of house dust mites (*Dermatophagoides* spp.) (Wassenaar, 1988), and that some water mites could reduce mosquito numbers in the field (Smith, 1988).

Recent or continuing systematic treatments of acarine groups of biocontrol interest, such as the Anystidae (Meyer and Ueckermann, 1987), Camerobiidae (Bolland, 1986), Cunaxidae (Smiley, unpublished), Erythraeidae (Southcott, 1972 and other papers), various gamasids (Karg, 1987 and other publications), Hemisarcoptidae (O'Connor and Houck, unpublished), Phytoseiidae (Chant, 1985 and continuing research; Schicha, 1987), Pionidae (Smith, 1976), several African raphignathoid families (Camerobiidae, Eupalopsellidae, Stigmaeidae) (Meyer and Ueckermann, 1989), Tarsonemidae (Lindquist, 1986) and exotic water mites (Cook, 1986 and earlier publications), will hopefully increase interest in and facilitate further research on the potential of promising members of these families. The lists of mite–pest associations published by several authors in Hoy *et al.* (1983), and the annotated compilation of the parasitic associations of water mite larvae with insect hosts (Smith and Oliver, 1986) should also encourage additional research.

This presentation includes 29 families. Part One begins with a key to acarine orders and suborders. This is followed by keys to families within each suborder, and by the 29 individual family chapters, which, for convenience, are presented in alphabetical order. Part Two consists of chapters of general interest (in part supplementing Hoy *et al.*, 1983) on the use of acarine biocontrol agents against nematodes, locusts, scale insects, aquatic Diptera of medical importance and stored-product pests; on the effects of the host-plant, and of pesticides, on pest control by Acari, on various methods for rearing and shipping acarine biocontrol agents; on methods of demonstrating the efficacy of mite predators and parasites, followed by a discussion of their attributes. Some general recommendations for further research conclude this work.

REFERENCES

Anonymous (1964) *Proc. 1st Int. Congr. Acarol.*, Fort Collins, Colorado, (USA) 2–7 September, 1963. *Acarologia*, **6**, Fasc. Hors Serie, pp. 1–439.

Axtell, R.C. (1969) Macrochelidae (Acarina: Mesostigmata) as biological control agents for synanthropic flies. In *Proceedings of the 2nd International Congress of Acarology*, (ed. G.O. Evans), Sutton Bonnington (England) 19–25 July, 1967. Akademiai Kiado, Budapest, pp. 401–16.

Axtell, R.C. and Rutz, D.A. (1986) Role of parasites and predators as biological fly control agents in poultry production facilities. *Misc. Publ. Entomol. Soc. Am.*, **61**, 88–100.

Binns, E.S. (1982) Phoresy as migration – some functional aspects of phoresy in mites. *Biol. Rev.*, **57**, 571–620.

Bolland, H.R. (1986) Review of the systematics of the family Camerobiidae (Acari, Raphignathoidea). 1. The genera *Camerobia, Decaphyllobious, Tillandsobius* and *Tycherobius. Tijd. Entomolo.*, **129**, 191–215.

Brust, G.E. and House, G.J. (1988) A study of *Tyrophagus putrescentiae* (Acari: Acaridae) as a facultative predator of southern corn rootworm eggs. *Exp. Appl. Acarol.*, **4**, 335–44.

Burnett, T. (1977) Biological models of two acarine predators of the grain mite, *Acarus siro* L. *Can. J. Zool.*, **55**, 1312–23.

Chant, D.A. (1985) Systematics and taxonomy. In *Spider Mites, their Biology, Natural Enemies and Control* (eds W. Helle and M. Sabelis), Elsevier, Amsterdam, vol. 1B, pp. 17–27.

Cook, D.R. (1986) Water mites from Australia. *Mem. Am. Entomol. Inst.*, **40**, 568.

Dicke, M. and Sabelis, M.W. (1988) Infochemical terminology: should it be based on cost-benefit analysis rather than origin of compounds? *Functional Ecol.*, **2**, 131–9.

Evans, G.O. (1969) *Proceedings of the 2nd International Congress of Acarology*, Sutton Bonnington (England) 19–25 July 1967. Akademiai Kiado, Budapest, p. 652.

Ewing, H.E. (1912) The life history and habits of *Cheyletus seminivorus* Packard. *J. Econ. Entomol.*, **5**, 416–20.

Gause, G.F., Smaragdova, N.P. and Witt, A.A. (1936) Further studies of interaction between predators and prey. *J. Anim. Ecol.*, **5**, 1–18.

Griffiths, D.A. and Bowman, C.E. (1984) *Acarology, VI*, Vol. 1–2. Ellis Horwood, Chichester.

Halliday, R.B. and Holm, E. (1987) Mites of the family Macrochelidae as predators of two species of dung-breeding pest flies. *Entomophaga*, **32**, 333–8.

Hansen, L.S. (1988) Control of *Thrips tabaci* (Thysanoptera: Thripidae) on glasshouse cucumber using large introductions of predatory mites *Amblyseius barkeri* (Acarina: Phytoseiidae). *Entomophaga*, **33**, 33–42.

Hessein, N.A. and Perring, T.M. (1986) Feeding habits of the Tydeidae with evidence of *Homeopronematus anconai* (Acari: Tydeidae) predation on *Aculops lycopersici* (Acari: Eriophyidae). *Int. J. Acarol.*, **12**, 215–21.

Hill, R.L. and Stone, C. (1985) Spider mites as control agents for weeds. In *Spider Mites: Their Biology, Natural Enemies and Control* (eds W. Helle and M. Sabelis), Elsevier, Amsterdam, vol. 1B, pp. 443–8.

Hochmuth, R.C., Hellman, J.L., Dively, G. and Schroder, R.F.W. (1987) Effect of the parasitic mite *Coccipolipus epilachnae* (Acari: Podapolipidae) on feeding, fecundity, and longevity of soybean-fed adult Mexican bean beetles (Coleoptera: Coccinellidae) at different temperatures. *J. Econ. Entomol.*, **80**, 612–16.

Holm, E. and Wallace, M.M.H. (1989) Distribution of some anystid mites (Acari: Anystidae) in Australia and Indonesia and their role as possible predators of the cattle tick, *Boophilus microplus* (Acari: Ixodidae). *Exp. Appl. Acarol.*, **6**, 77–88.

Howard, L.O. (1930) *A History of Applied Entomology (Somewhat Anecdotal)*. Smithsonian Misc. Coll., Publ. No. 3065, p. 564.

Howard, L.O., Dyar, H.G. and Knab, F. (1912) *The Mosquitoes of North and Central America and the West Indies. (General Considerations of Mosquitoes, Their Habits and Their Relation to the Human Species*, vol. 1), Carnegie Institution, Washington, 520 pp.

Hoy, M.A. (ed.) (1982) *Recent Advances in Knowledge of the Phytoseiidae*. Division of Agricultural Science, University of California, Pub. no. 3284, p. 92.

Hoy, M.A., Cunningham, G.L. and Knutson, L. (eds) (1983) *Biological Control of Pests by Mites*. University of California, Special Publication no. 3304, 185 pp.

Kaiser, H. (1983) Small scale spatial heterogeneity influences predation success in an unexpected way: model experiments on the functional response of predatory mites (Acarina). *Oecologia*, **56**, 249–56.

Karg, W. (1987) Neue Raubmilbenarten der Gattung *Hypoaspis* Canestrini, 1884 (Acarina: Parasitiformes). *Zool. Jb. Syst.*, **114**, 289–302.

Kinn, D.N. (1983) Mites as biological agents of bark and sawyer beetles. In *Biological Control of Pests by Mites* (eds M.A. Hoy, G.L. Cunningham and L. Knutson), University of California special publication, no. 3304, p. 67–73.

Lindquist, E.E. (1969) Mites and the regulation of bark beetle populations. In *Proceedings of the 2nd International Congress of Acarology*, Sutton Bonnington (England) 19–25 July, 1967 (ed. G.O. Evans), Akademiai Kiado, Budapest, pp. 389–99.

Lindquist, E.E. (1986) The world genera of Tarsonemidae (Acari: Heterostigmata): a morphological, phylogenetic and systematic revision, with a reclassification of family-group taxa in the Heterostigmata. *Mem. Entomol. Soc. Can.*, **136**, 1–517.

Logan, P.A., Casagrande, R.A., Hsiao, T.H. and Drummond, F.A. (1987) Collections of natural enemies of *Leptinotarsa decemlineata* (Coleoptera: Chrysomelidae) in Mexico, 1980–1985. *Entomophaga*, **32**, 249–54.

McMurtry, J.A. and Rodriguez, J.G. (1987) Nutritional ecology of phytoseiid mites. In *Nutritional Ecology of Insects, Mites and Spiders* (eds F. Slansky Jr and J.G. Rodriguez), Wiley, Chichester, pp. 609–44.

Megevand, B., Yaninek, J.S. and Friese, D.D. (1987) Classical biological control of the cassava green mite. *Insect Sc. Appl.*, **8**, 871–4.

Meyer, M.K.P. (Smith) and Ueckermann, E.A. (1987) A taxonomic study of some Anystidae (Acari: Prostigmata). *Entomol. Mem. Dept. Agric. Wat. Supp. South Africa*, **68**, 1–37.

Meyer, M.K.P. (Smith) and Ueckermann, E.A. (1989) African Raphignathoidea

(Acari: Prostigmata). *Entomol. Mem. Dept. Agric. Wat. Supp. South Africa*, **74**, 1–58.

Meyerdirk, D.E. and Coudriet, D.L. (1985) Predation and developmental studies of *Euseius hibisci* (Chant) (Acarina: Phytoseiidae) feeding on *Bemisia tabaci* (Gennadius) (Homoptera: Aleyrodidae). *Environ. Entomol.*, **14**, 24–7.

Moser, J.C. (1975) Mite predators of the southern pine beetle. *Ann. Entomol. Soc. Am.*, **68**, 1113–16.

Nachman, G. (1987) Systems analysis of acarine predator–prey interactions. I. A stochastic simulation model of spatial processes. *J. Anim. Ecol.*, **56**, 247–65.

Ramakers, P.M.J. and van Lieburg, M.J. (1982) Start of commercial production and introduction of *Amblyseius mckenziei* Sch. and Pr. (Acarina: Phytoseiidae) for the control of *Thrips tabaci* Lind. (Thysanoptera: Thripidae) in glasshouses. *Med. Fac. Landbouww. Rijksuniv. Gent*, **47/2**, 541–5.

Rangel, A.F. (1901) Cuarto informe acerca del picudo del algodon (*Insanthonomus grandis* I. C. Cu.). *Bol. Comis. Parasitol. Agri.*, **1**, 245–61.

Riley, C.V. (1873) *Fifth Annual Report on the Noxious, Beneficial and Other Insects of the State of Missouri*, Regan and Carter, p. 87.

Riley, C.V. (1874a) Descriptions of two new subterranean mites. *Trans. Acad. Sci. St Louis*, **25**, 215–16.

Riley, C.V. (1874b) *Sixth Annual Report on the Noxius, Beneficial and Other Insects of the State of Missouri*, Regan and Carter, p. 52.

Riley, C.V. (1878) Egg-feeding mites. *Can. Entomol.*, **10**, 58–9.

Rust, H.J. (1933) Many bark beetles destroyed by predacious mite. *J. Econ. Entomol.*, **26**, 733–4.

Sabelis, M.W. (1985) Predation on spider mites. In *Spider Mites, Their Biology, Natural Enemies and Control* (eds W. Helle and M. Sabelis), Elsevier, Amsterdam, Vol. 1B, pp. 103–29.

Saito, Y. (1986) Prey kills predators: counter attack success of a spider mite against its specific phytoseiid predator. *Exp. Appl. Acarol.*, **2**, 47–62.

Saito, Y. (1987) Extraordinary effects of fertilization status on the reproduction of an arrhenotokous and sub-social spider mite (Acari: Tetranychidae). *Res. Popul. Ecol.*, **29**, 57–71.

Schicha, E. (1987) *Phytoseiidae of Australia and Neighboring Areas*, Indira Publishing House, Oak Park, Maryland, p. 187.

Shimer, H. (1868a) Notes on the 'apple bark-louse' (*Lepidosaphes conchiformis*, Gmelin sp.) with a description of a supposed new Acarus. *Trans. Am. Entomol. Soc.*, **1**, 361–74.

Shimer, H. (1868b) The apple bark louse in 1866. Birds vindicated from the charge preferred against them by the State Entomologist. *Trans. Illinois St Hort. Soc.*, n.s. **2**, 227–33.

Smith, B.P. (1988) Host–parasite interaction and impact of larval water mites on insects. *Ann. Rev. Entomol.*, **33**, 487–507.

Smith, I.M. (1976) A study of the systematics of the water mite family Pionidae Prostigmata: Parasitengona). *Mem. Entomol. Soc. Can.*, **98**, 1–249.

Smith, I.M. and Oliver, D.R. (1986) Review of parasitic associations of larval water mites (Acari: Parasitengona: Hydrachnida) with insect hosts. *Can. Entomol.*, **118**, 407–72.

Solomon, M.E. (1946) Tyroglyphid mites in stored products. Ecological studies. *Ann. Appl. Biol.*, **33**, 82–97.

Southcott, R.V. (1972) Revision of the larvae of the tribe Callidosomatini (Acarina: Erythraeidae) with observations on post-larval instars. *Aust. J. Zool., Suppl. Ser.*, **13**, 1–84.

Tanigoshi, L.K., Nishio-Wong, J.Y. and Fargerlund, J. (1984) *Euseius hibisci*: Its control of citrus thrips in southern California citrus orchards. In *Acarology VI* (eds D.A. Griffiths and C.E. Bowman), Ellis Horwood, Chichester, **2**, 699–702.

Thorvilson, H.G., Phillips, S.A. Jr, Sorensen, A.A. and Trostle, M.R. (1987) The straw itch mite, *Pyemotes tritici* (Acari: Pyemotidae), as a biological control agent of red imported fire ants, *Solenopsis invicta* (Hymenoptera: Formicidae). *Fla. Entomol.*, **70**, 439–44.

Tothill, J.D. (1918) Natural control investigations in British Columbia. *Proc. Entomol. Soc. British Columbia*, **12**, 37–9.

Turnbull, A.L. and Chant, D.A. (1961) The practice and theory of biological control in Canada. *Can. J. Zool.*, **39**, 697–753.

Uchida, T. and Miyazaki, I. (1935) Life history of a water-mite parasitic on *Anopheles*. *Proc. Imp. Acad. Japan*, **11**, 73–6.

Wallace, M.M.H. and Walters, M.C. (1974) The introduction of *Bdellodes lapidaria* (Acari: Bdellidae) from Australia into South Africa for the biological control of *Sminthurus viridis* (Collembola). *Aust. J. Zool.*, **22**, 505–17.

Walter, D.E., Hunt, H.W. and Elliot, E.T. (1988) Guilds or functional groups? An analysis of predatory arthropods from a shortgrass steppe soil. *Pedobiologia*, **31**, 247–60.

Wassenaar, D.P.J. (1988) Effectiveness of vacuum cleaning and wet cleaning in reducing house-dust mites, fungi and mite allergen in a cotton carpet: a case study. *Exp. Appl. Acarol.*, **4**, 53–62.

Webster, F.M. (1910) A predaceous and supposedly beneficial mite, *Pediculoides*, becomes noxious to man. *Ann. Entomol. Soc. Am.*, **3**, 15–39.

Womersley, H. (1933) A possible biological control of the clover springtail or lucerne flea (*Sminthurus viridis* L.) in Western Australia. *J. Aust. Counc. Sci. Indust. Res.*, **6**, 83–91.

Žďárková, E. (1986) Mass rearing of the predator *Cheyletus eruditus* (Schrank) (Acarina: Cheyletidae) for biological control of acarid mites infesting stored products. *Crop Prot.*, **5**, 122–4.

PART ONE
Illustrated Key and Family Descriptions

2

Illustrated key to relevant acarine families

The characters of the adult female mite form the basis for most of the key couplets. Occasional references to characters of the hypopodes (formerly termed hypopi) and larval water mites are provided merely as supplementary information. The hypopus (deutonymph or second nymphal instar) occurs exclusively in the suborder Astigmata, and then only occasionally. It differs widely from preceding and following developmental stages. Zakhvatkin (1959) should be consulted to keys for hypopodes. Parasitic larval water mites often show totally different facies than their later stages.

The abdominal segmentation of spiders, mites and ticks is inconspicuous or apparently absent. In the subclass Araneae (spiders), the head and thorax are combined in a single unit, the cephalothorax, joined to the abdomen by a slender pedicel. The subclass Acari includes the ticks (Fig. 2) and the mites (Fig. 3). The acarine body, in contrast to the rather distinctly divided body of insects, is composed mainly of the idiosoma. The mouthparts are borne on the gnathosoma, an anterior region more or less distinct from the idiosoma. Terminology for the subdivisions of the acarine body are given in Fig. 1.

The following key, which is based on Baker and Wharton (1952), Cook (1974), Krantz (1978), Mullen (1974) and Smiley (in press), begins with the relevant orders and continues to the suborders. Several hierarchial systems of these taxa are currently employed by acarologists, and Table 1 shows some of these systems along with one used herein.

The main objective of the key is to help users identify families of mites currently known to be of potential or realized biocontrol importance. Although family recognition is emphasized, we have included some extra information on morphological structures to identify mites at the ordinal and subordinal levels, in the hope that additional mite families of relevant interest will be discovered. Alphanumerics in the following section (1A, 5D etc) refer to the included plates. Also included are some scanning electron micrographs of mite families as well as photographs of damage inflicted by the mites on pest or weed prey and hosts.

HEMISARCOPTIDAE

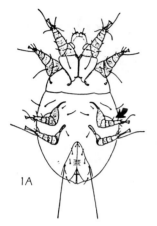

Hemisarcoptes malus
(After Baker and Wharton, 1952)

Key to orders

1. Without conspicuous lateroventral hysterosomal stigmata (Fig. 1A) (cryptic stigmata present in some Cryptostigmata); sensilla (Figs 1B, 1C and 1F) present or absent ..... Order Acariformes 2
   With 1–4 conspicuous lateroventral hysterosomal stigmata (Figs 1D and 1E); sensilla absent (Fig. 1G) ..... Order Parasitiformes 26

GALUMNIDAE

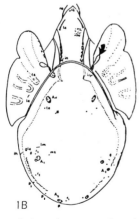

Orthogalumna terebrantis

PYEMOTIDAE

Pyemotes tritici

ASCIDAE

TROMBIDIIDAE

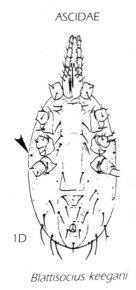

1D *Blattisocius keegani*

1F trombiid mite
(After Krantz, 1978)

UROPODIDAE

ASCIDAE

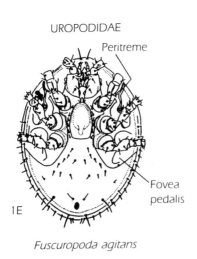

1E *Fuscuropoda agitans*

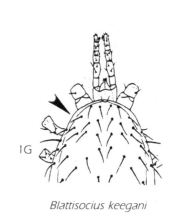

1G *Blattisocius keegani*

Table 1 Names proposed for the higher acarine taxa (orders and suborders). Several systems were proposed for acarine orders and suborders. This table makes a comparison between some of these systems. System I: Baker and Wharton (1952); System II: Evans et al. (1961); System III: Krantz (1978); System IV: Smiley (in press) and in the present book.

System I	System II	System III	System IV
Order Parasitiformes			
Onychopalpida			
Holothyroidea	Tetrastigmata	Holothyrida	Holothyrina*
Notostigmata	Notostigmata	Opilioacarida	Opilioacarida*
Parasitiformes			
Ixodides	Metastigmata	Ixodida	Ixodida†
Mesostigmata	Mesostigmata	Gamasida	Mesostigmata
Order Acariformes			
Trombidiformes	Prostigmata	Actineida	Prostigmata
Tetrapodili			Eriophyoidea
Sarcoptiformes			
Acaridiae	Astigmata	Acaridida	Astigmata
Oribatei	Cryptostigmata	Oribatida	Cryptostigmata

* It is unlikely that users of this manual will ever encounter mites belonging to either of these rare and exotic suborders.
† Ticks are included here only to show their position in acarine classification.

ORDER: ACARIFORMES

Key to suborders

2. Stigmata absent or inconspicuous; in later case, never opening between the bases of the chelicerae or on the anterior shoulders of the propodosoma ..... Suborders Cryptostigmata and Astigmata 3

Since stigmata are always difficult to see in the Cryptostigmata (hence the name of the suborder), and are sometimes hard to see in the Prostigmata,

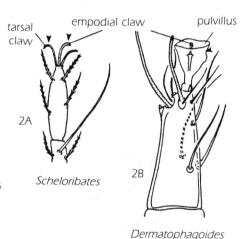

ORIBATULIDAE   PYROGLYPHIDAE

2A *Scheloribates*

2B *Dermatophagoides pteronyssinus*

TYDEIDAE

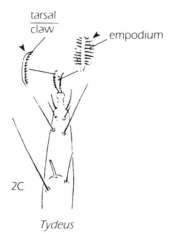

Tydeus

CHEYLETIDAE

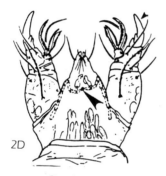

Cheyletus eruditus

HEMISARCOPTIDAE

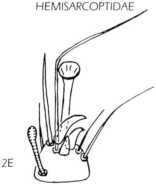

Hemisarcoptes malus

additional characters are offered to the reader in order to separate the Cryptostigmata and Astigmata from the Prostigmata: Tarsal claw present, without tenent hairs (Figs 2A, 2B); empodium usually present, often claw-like (Fig. 2A) (claw may be very small (Fig. 2B) or absent from pulvillus (Fig. 2E) and sometimes surrounded by a sucker-like pulvillus (Fig. 2B)); chelicerae seldom hooked or styletiform (Fig. 2K), usually strong, chelate, and dentate (Fig. 2F and 2K); palpus sometimes with retrorse teeth, but never with a thumb-claw complex or strong claw (Fig. 2G); ventral opisthosoma sometimes with claspers or anal plate suckers (Fig. 2H); hypopodes may be present (Fig. 2I).

— Stigmata present and opening either between the bases of the chelicerae or on the anterior shoulders of the propodosoma ..... Suborder Prostigmata 6

Tarsal claws, if present, sometimes with tenent hairs (Fig. 2C); empodium, if present, usually claw-like or pad-like (Fig. 2C), and not surrounded by a suckerlike pulvillus; chelicerae usually hook-like (Fig. 2J) or styletiform (Fig. 2L), rarely strong, chelate (Fig. 2G) and dentate; palpus without retrorse teeth, but may be with a strong claw and/or thumb-claw complex (Fig. 2D); ventral opisthosoma without claspers or anal plate suckers (Fig. 2M); no hypopodes.

GLYCYPHAGIDAE

2F

Lepidoglyphus destructor

CARPOGLYPHIDAE

Carpoglyphus lactis

2G

ACARIDAE

Suidasia pontifica

2H

ACARIDAE

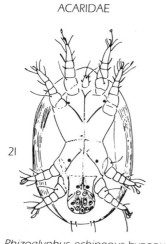

2I

Rhizoglyphus echinopus hypopus

ANYSTIDAE

2J

Anystis

PTERYGOSOMATIDAE

2K

Geckobiella texana
(After Krantz, 1978)

PYEMOTIDAE

2L

Pyemotes tritici

BDELLIDAE

2M

eye

Spinibdella bifurcata

ORIBATULIDAE

3A

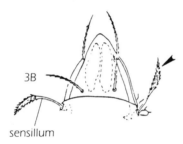

3B

sensillum

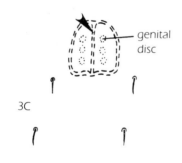

genital disc

3C

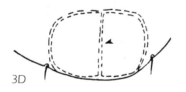

3D

Scheloribates (A–D)

3. Sclerotization of adult integument usually strongly developed, body brown or black; genital orifice longitudinal (Fig. 3C), usually flanked by three pairs of discs; genital (Fig. 3C) and anal (Fig. 3D) orifices similar in shape and covered by trapdoor-like valves; sexes usually homomorphic; empodium, if present, claw-like, not borne on pretarsus (Fig. 3A); sensory setae (sensilla or trichobothria) (Fig. 3B) located on propodosoma ..... Suborder Cryptostigmata 4

— Sclerotization of adult integument weak or lacking, body whitish; genital orifice transverse or U-, V-, or Y-shaped (Figs 3H and 3I), usually flanked by two pairs of genital discs; genital (Figs 3H and 3I) and anal (Figs 3J and 3K) orifices dissimilar, sexes homomorphic or heteromorphic; empodium claw-like (Figs 3E and 3F), often borne on pretarsus; no sensory setae on propodosoma (Fig. 3G) ..... Suborder Astigmata .. 5

PYROGLYPHIDAE

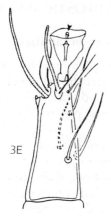

3E

Dermatophagoides pteronyssinus

GLYCYPHAGIDAE

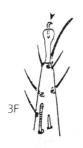

3F

Gohieria fusca

ACARIDAE

3G

Tyrophagus putrescentiae

PYROGLYPHIDAE

3H

CARPOGLYPHIDAE

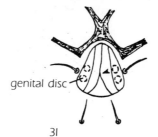

genital disc

3I

3J

Dermatophagoides farinae (H and J)

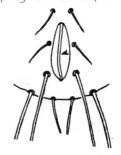

3K

Carpoglyphus lactis (I and K)

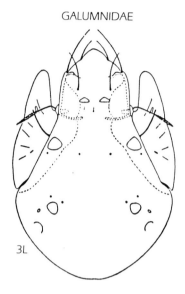

GALUMNIDAE

3L

SUBORDER: CRYPTOSTIGMATA

Key to families

4. With pteromorphae (wing-like structures) which are rounded anteriorly and pointed posteriorly .... GALUMNIDAE
— With or without pteromorphae; if with these structures, they are never rounded anteriorly and pointed posteriorly .... other Cryptostigmata

The only cryptostigmatid family to be of any significance in pest biocontrol is the Galumnidae (Figs 3L, 3M and 3N).

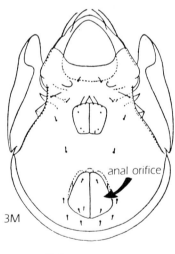

Galumna virginiensis
(After Baker and Wharton, 1952)

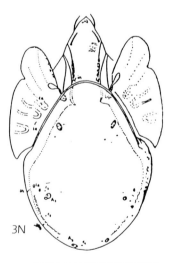

Orthogalumna terebrantis

ACARIDAE

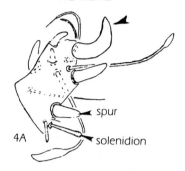

Rhizoglyphus robini

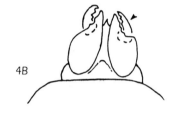

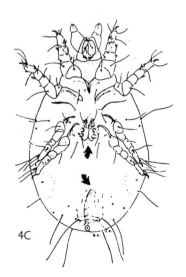

Rhizoglyphus echinopus
(After Manson, 1972)

SUBORDER: ASTIGMATA

Key to families

5. Empodial claw (Fig. 4A) large and distinct; chelicerae chelate-dentate (Fig. 4B); genital and anal plates separated (Fig. 4C); genital plate never located below level of coxae IV; setae sce and sci (Fig. 4D), when present, inserted on a horizontal plane
.... ACARIDAE

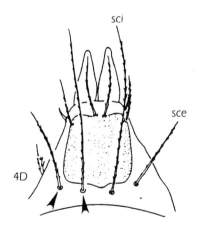

Tyrophagus putrescentiae

— Empodial claw absent (Fig. 5A); chelicerae chelate but not denticulate (Fig. 5B); genital and anal shields confluent (Fig. 5D), located on a level between and behind coxae IV; seta sce and sci (Fig. 5C) located on a vertical plane
.... HEMISARCOPTIDAE

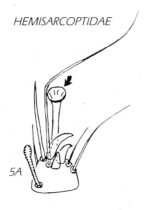

Hemisarcoptes malus

Hemisarcoptes

Hemisarcoptes coccophagus
(After Gerson and Schneider, 1981)

Hemisarcoptes malus
(After Baker and Wharton, 1952)

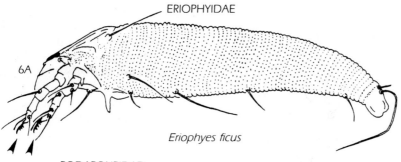

Eriophyes ficus

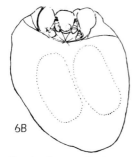

Coccipolipus epilachnae
adult female

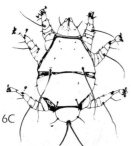

Coccipolipus epilachnae
larviform female, dorsum

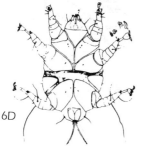

Coccipolipus epilachnae
larviform female, venter

SUBORDER: PROSTIGMATA

Key to families

6. Found in aquatic habitats and often capable of swimming 22
— Rarely found in aquatic habitats; terrestrial free-living predators or parasites of animals or plants 7

7. With two (Fig. 6A and 6B) or three (Figs 6C and 6D) pairs of legs 8
— With four (Figs 8A–E) pairs of legs 9

8. Body worm-like (Fig. 6A); with two pairs of legs .... ERIOPHYIDAE

This family represents the phytophagous superfamily Eriophyoidea, minute mites variously called bud, gall, rust, or blister mites dependent on their location and the kind of reaction they cause in the host plant.

— Body (Fig. 6B) sac-like or elongate (Figs 6C and 6D), but not worm-like; with one to three pairs of legs .... PODAPOLIPIDAE

9. Without (Fig. 7A) a palpal thumb-claw process 10
— With (Fig. 7B) a distinct palpal claw or thumb-claw process (Fig. 7C) 14

TYDEIDAE

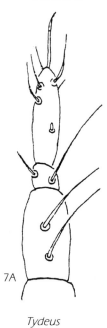

7A

Tydeus

TETRANYCHIDAE

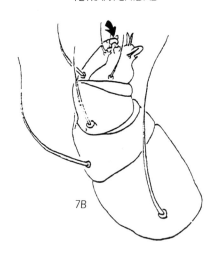

7B

Bryobia praetiosa

CHEYLETIDAE

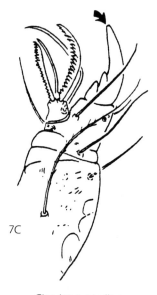

7C

Cheyletus eruditus

BDELLIDAE

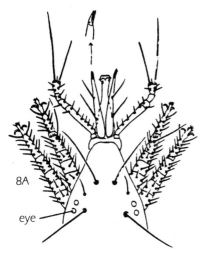

Spinibdella bifurcata
(After Baker and Balock, 1944)

BDELLIDAE

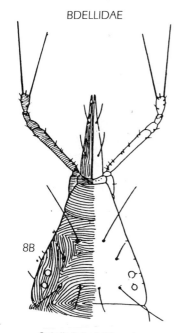

Spinibdella birfurcata
(After Atyeo, 1960)

10. Cheliceral bases not fused; chelicerae hinged at base, moving scissor-like over gnathosoma (Figs 8A–C); propodosoma with two pairs of long sensory setae or sensilla (Figs 8A–C) 11
— Cheliceral bases fused (Fig. 8E); if not fused, not capable of lateral scissor-like motion (Fig. 8E); propodosoma with one pair of sensory setae or sensilla (Fig. 8E) 12

11. Gnathosoma snout-like (Fig. 8A); palpus without multi-branched setae spines, spurs, or apophyses (Figs 8A–B), terminating with strong setae .... BDELLIDAE
— Gnathosoma cone-like (Fig. 8C); palpus with or without multi-branched setae (Fig. 8C), with either spurs, spines, or apophyses, terminating with a claw (all genera except *Parabonzia*) (Fig. 8D) .... CUNAXIDAE

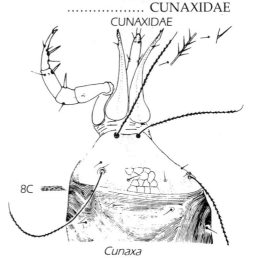

Cunaxa

27

CUNAXIDAE

8D

Parabonzia bdelliformis

TYDEIDAE

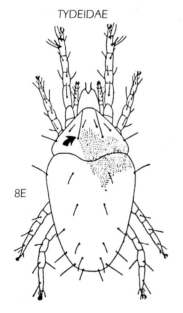

8E

Tydeus starri
(After Baker and Wharton, 1952)

TYDEIDAE

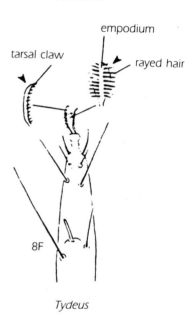

8F

Tydeus

12. Sensillum (prodorsal trichobothrium) of female globular (Figs 9A–C); tarsal claw and pulvillus smooth, without rayed hairs (Figs 9A–C) 13
— Sensillum of female setaceous (Fig. 8E); tarsal claw and empodium either smooth or pectinate, rayed hairs present (Fig. 8F) .... TYDEIDAE

13. Leg IV of female with pretarsus, claw and pulvillus, but without apical whip-like setae (Figs 9A – 9B); idiosoma elongate (Fig. 9A); gravid female with sac-like hysterosoma (Fig. 9B) .... PYEMOTIDAE
— Leg IV of female without pretarsus, claw and pulvillus, but with apical whip-like setae (Fig. 9C); idiosoma oval (Fig. 9C); gravid female without sac-like hysterosoma .... TARSONEMIDAE

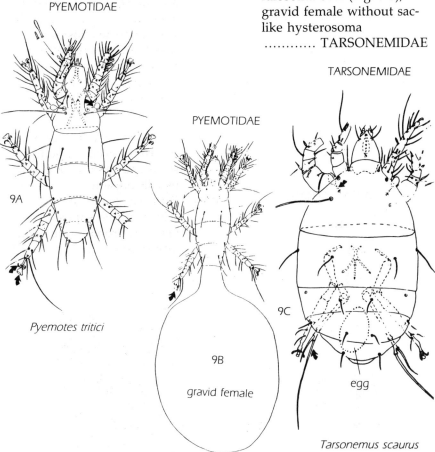

PYEMOTIDAE

PYEMOTIDAE

TARSONEMIDAE

9A

Pyemotes tritici

9B

gravid female

Pyemotes tritici

9C

egg

Tarsonemus scaurus

PTERYGOSOMATIDAE

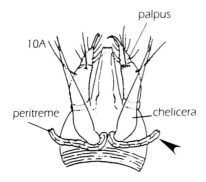

Pimeliaphilus cunliffei

TETRANYCHIDAE

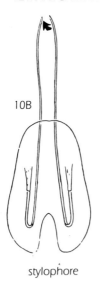

stylophore

14. Chelicerae not whip-like (Figs 10A and 10C) and not arising from eversible stylophore 15
— Chelicerae fused and whip-like (Figs 10B and 10E) and arising from eversible stylophore
   .... TETRANYCHIDAE

15. Chelicerae and rostrum not fused into a cone and peritremes not forming an arch of M-shaped configuration 16
— Chelicerae and rostrum fused into a cone, peritremes arched or forming an M-shaped configuration (Figs 10C and 10D)
   .... CHEYLETIDAE

16. Peritremes not emergent; suture between propodosoma and hysterosoma usually present 18
— Peritremes emergent (Fig. 10A); no suture between propodosoma and hysterosoma 17

CHEYLETIDAE

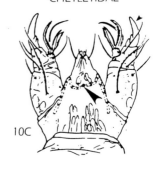

Cheyletus eruditus

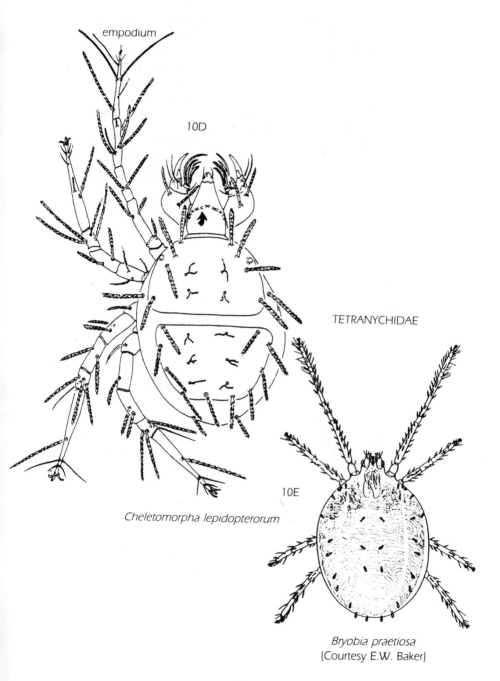

CHEYLETIDAE

empodium

10D

TETRANYCHIDAE

Cheletomorpha lepidopterorum

10E

Bryobia praetiosa
(Courtesy E.W. Baker)

ANYSTIDAE

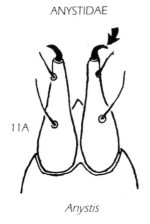

Anystis

17. Palpal tibiae (Fig. 11B) with claw; chelicerae strongly hooked distally (Fig. 11A); tarsi with two claws, which may be combed, toothed or pilose and with a claw-like, brush-like, or bell-like empodium .... ANYSTIDAE
— Palpal tibiae without claws, chelicerae chelate and weakly developed (Fig. 11C); tarsi with two claws with tenent hairs but without empodia .... PTERYGOSOMATIDAE

ANYSTIDAE

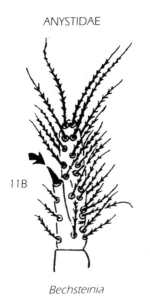

Bechsteinia

PTERYGOSOMATIDAE

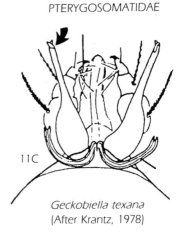

Geckobiella texana
(After Krantz, 1978)

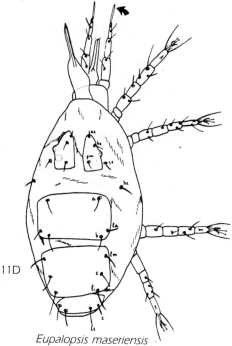

EUPALOPSELLIDAE

11D

Eupalopsis maseriensis
(After Gerson, 1966)

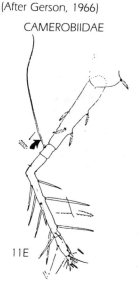

CAMEROBIIDAE

11E

Neophyllobius Iorioi

18. Dorsal propodosoma without crista metopica and sensilla (Figs 11D, 11F and 12A) 19
— Dorsal propodosoma with crista metopica and sensilla (Figs 12B and 12C) 21

19. Palpal claw distinct (Figs 12A–C) 20
— Palpal claw reduced (Fig. 11D); gnathosoma and chelicerae both elongate (Fig. 11D)
.... EUPALOPSELLIDAE

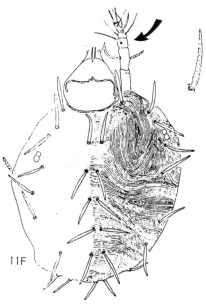

CAMEROBIIDAE

11F

Neophyllobius Iorioi

STIGMAEIDAE

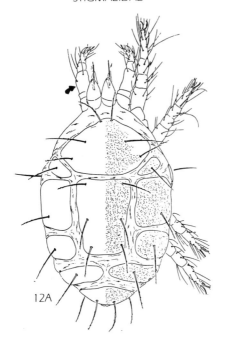

Stigmaeus glypticus
(After Summers, 1962)

20. Palpi weak (Fig. 11F); legs stilted, longer than body; genua I–IV usually with extremely long setae (Fig. 11E)
.... CAMEROBIIDAE
— Palpi robust (Fig. 12A); legs not longer than body; genua I–IV without extremely long setae (Fig. 12A)
.... STIGMAEIDAE

21. Chelicerae long (Fig. 12C), retractable into region of hysterosoma with two pairs of prodorsal trichobothria (Fig. 12C)
.... ERYTHRAEIDAE
— Chelicerae short (Fig. 12B), non-retractable into region of hysterosoma; with one pair of prodorsal trichobothria (Fig. 12B)
.... TROMBIDIIDAE

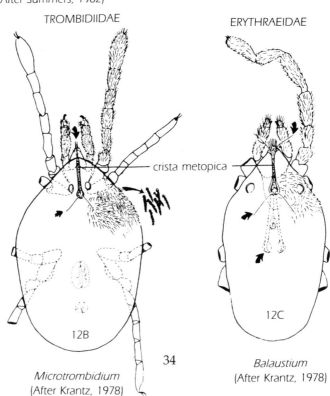

TROMBIDIIDAE

Microtrombidium
(After Krantz, 1978)

ERYTHRAEIDAE

Balaustium
(After Krantz, 1978)

ARRENURIDAE

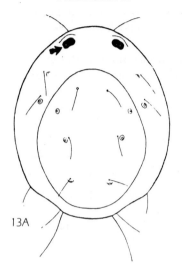

13A

Arrenurus (Megaluracarus) liberiensis

LIMNESIIDAE

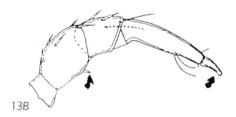

13B

Limnesia lucifera uniseta

HYDRYPHANTIDAE

13C

Thyas stolli

22. Lateral eyes in distinct capsules and these lying on the soft integument (Fig. 13C); palp chelate, i.e. distal end of tibia extending beyond insertion of tarsus (Fig. 13D) ....
HYDRYPHANTIDAE
(*Thyas, Thyasides*)
— Lateral eyes not in distinct capsules or capsules incorporated into the heavily sclerotized dorsum (Fig. 13A); palp not chelate (Fig. 13B) 23

HYDRYPHANTIDAE

13D

Deuterothyas variabilis

LIMNESIIDAE

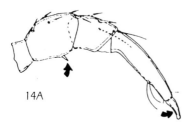

14A

Limnesia lembangensis

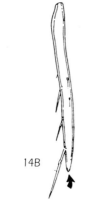

14B

Limnesia (Tetralimnesia) pinguipalpis

23. Integument weak; palpal telofemur with a single peg-like or hair-like seta on ventral side (Fig. 14A); no claws on the fourth legs (Fig. 14B) .... LIMNESIIDAE
(*Limnesia*)
— Integument either weak or heavily sclerotized; palpal telofemur without a ventral seta; fourth legs with well-developed claws 24

24. Body heavily sclerotized, with closely fitting dorsal (Fig. 15A) and ventral shields (Fig. 15B) present
.... 25
— Body usually soft, rarely males may have large dorsal and ventral sclerites but these not forming closely fitting dorsal and ventral shields (Fig. 15C)
.... PIONIDAE
(*Piona*)

ARRENURIDAE

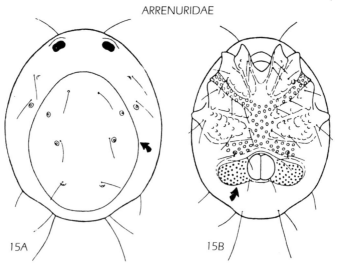

Arrenurus (Megaluracarus) liberiensis

PIONIDAE

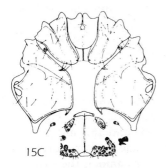

Piona catatama

ARRENURIDAE

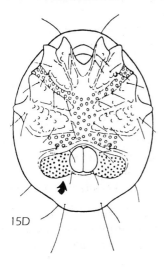

Arrenurus (Megaluracarus) liberiensis

ARRENURIDAE

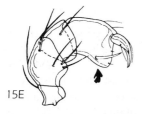

Arrenurus pseudoaffinis

25. Genital acetabula on wing-like acetabular plates which are incorporated into the ventral shield (Fig. 15D); ventral side of palpal tibia bulging to form an uncate palp (Fig. 15E)
.... ARRENURIDE
(*Arrenurus*)

— Genital acetabula lying free in the gonopore (Fig. 15F); ventral side of palpal tibia usually not bulging (Fig. 15G) but rarely may be as indicated in figure (Fig. 15E)
.... MIDEOPSIDAE
(*Mideopsis*)

MIDEOPSIDAE

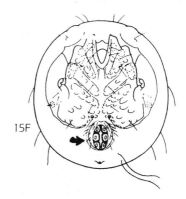

Mideopsis fibrosa

MIDEOPSIDAE

Mideopsis fibrosa

ARRENURIDAE

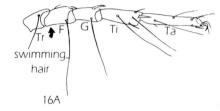

16A

HYDRYPHANTIDAE

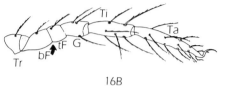

16B

MIDEOPSIDAE   ARRENURIDAE

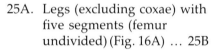

16C   16D

Mideopsis   *Arrenurus*

LIMNESIIDAE   PIONIDAE

16E   16F

Limnesia   *Piona*

[The family Mideopsidae is not included in the text. Figures 15D, 15F, 16C and 16D are provided to show some similar morphological structures among the genera.]

The last four families are water mites, whose parasitic larvae are the stage most often encountered in biocontrol situations. A key to the larvae of these four families (adapted from Mullen, 1974) is therefore enclosed.

25A. Legs (excluding coxae) with five segments (femur undivided) (Fig. 16A) ... 25B

— Legs (excluding coxae) with six segments (femur subdivided) (Fig. 16B) .... HYDRYPHANTIDAE

25B. Coxal plates fused together, but suture lines remaining between coxae (Fig. 16C) 25C

— Coxal plates separated from each other by a medial membraneous area (Fig. 16D) .... ARRENURIDAE

25C. Coxal plates 1–3 fused; suture lines present (Fig. 16E) .... LIMNESIIDAE

— Coxal plate 1 clearly separated from fused coxal plates 2 and 3 (Fig. 16F) .... PIONIDAE

ORDER: PARASITIFORMES

Key to suborders

26. Large leathery Acari (usually > 1 mm long), with sclerotized plates or shields (e.g. scutum, Fig. 17E); palpus without tined apotele (Fig. 17A); hypostome with retrorse teeth (Fig. 17A); spiracular (stigmal) plates (surrounding the stigma) oval, rounded, comma-shaped, or sub-triangular (Fig. 17E), and located between coxae III–IV, or behind coxae IV; Haller's organ (on dorsum of tarsus I) present (Fig. 17B); external parasites of vertebrates ... Suborder Ixodides (ticks)

17A

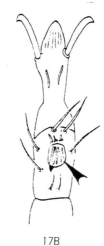

17B

Ticks, which parasitize vertebrates, are not known to affect agricultural pests. However, the occasional parasitism of female ticks by conspecific and other males (Oliver *et al.*, 1986) and tick attack of horse flies (Tabanidae) (Boshko and Skylar, 1981) have been recorded. Ticks were also among arthropods listed (in the role of vectors of various diseases) for biological warfare (Lockwood, 1987); their use in the control of vertebrate pests could therefore be considered.

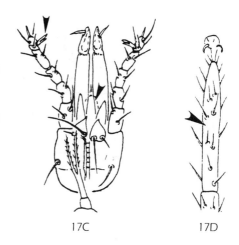

17C   17D

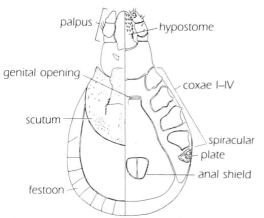

— Small non-leathery Acari (usually < 1 mm long); with or without sclerotized plates or shields; palpus with tined apotele (Fig. 17C); hypostome without retrorse teeth (Fig. 17C); stigma located between coxae II and III or III and IV, and sometimes surrounded by an elongate peritremal shield (Fig. 17F); Haller's organ absent (Fig. 17D); parasitic or free-living
..... Suborder Mesostigmata
.... 27

17E Tick, Suborder Ixodida (diagrammatic). Left, dorsal; right, ventral.

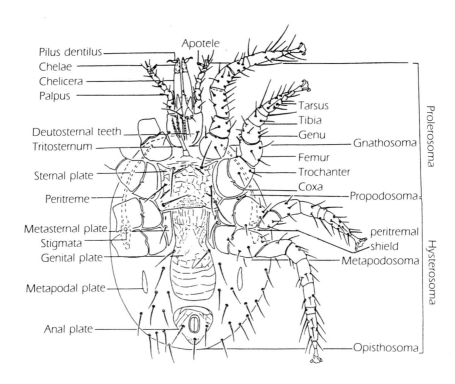

17F Tick Suborder Mesostigmata, LAELAPIDAE, *Androlaelaps casalis*

ASCIDAE

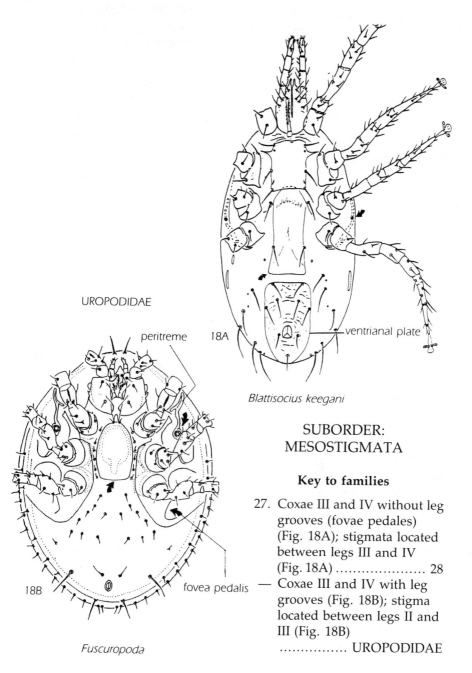

Blattisocius keegani

UROPODIDAE

Fuscuropoda

SUBORDER: MESOSTIGMATA

Key to families

27. Coxae III and IV without leg grooves (fovae pedales) (Fig. 18A); stigmata located between legs III and IV (Fig. 18A) 28
— Coxae III and IV with leg grooves (Fig. 18B); stigma located between legs II and III (Fig. 18B)
.... UROPODIDAE

ASCIDAE

19A

Blattisocius keegani

MACROCHELIDAE

19B

Macrocheles muscaedomesticae

28. Epigynial shield or genital plate not triangular (Figs 18A, 19D, 20A and 20B) 29
— Epigynial shield or genital plate triangular (Fig. 19C) .... PARASITIDAE

29. Leg I with ambulacra (claws and empodium) (Fig. 19A); peritremes not looped around stigmata (Figs 19C and 19D) 30
— Leg I without ambulacra (Fig. 19B); peritremes looped around stigmata (Fig. 20A) .... MACROCHELIDAE

ASCIDAE

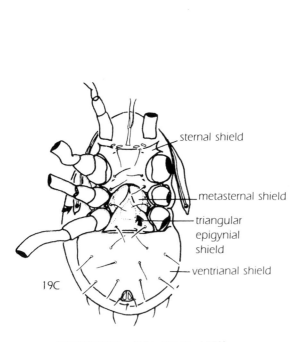

PARASITIDAE (After Krantz, 1978)

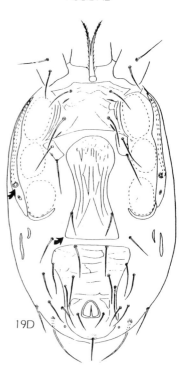

Blattisocius

30. Epigynial shield truncate posteriorly and about equal in length to ventrianal shield (Fig. 19D) 31
— Epigynial shield truncate posteriorly and at least twice as long as ventrianal shield (Fig. 20B) .... LAELAPIDAE

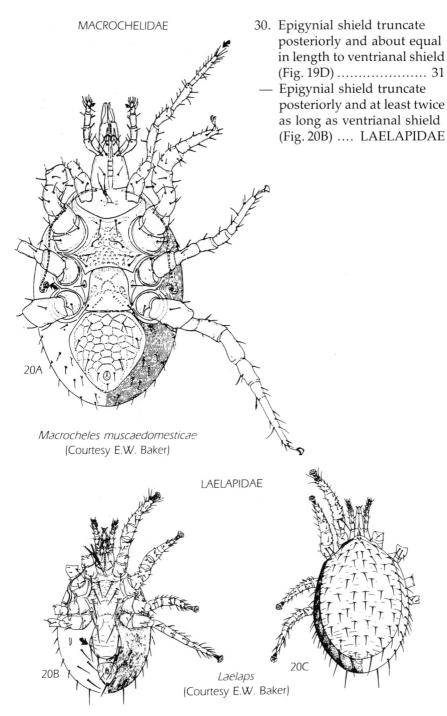

MACROCHELIDAE

Macrocheles muscaedomesticae
(Courtesy E.W. Baker)

LAELAPIDAE

Laelaps
(Courtesy E.W. Baker)

31. Dorsal shield with more than 20 pairs of setae (Fig. 21) .... ASCIDAE
— Dorsal shield with less than 20 pairs of setae (Fig. 22) .... PHYTOSEIIDAE

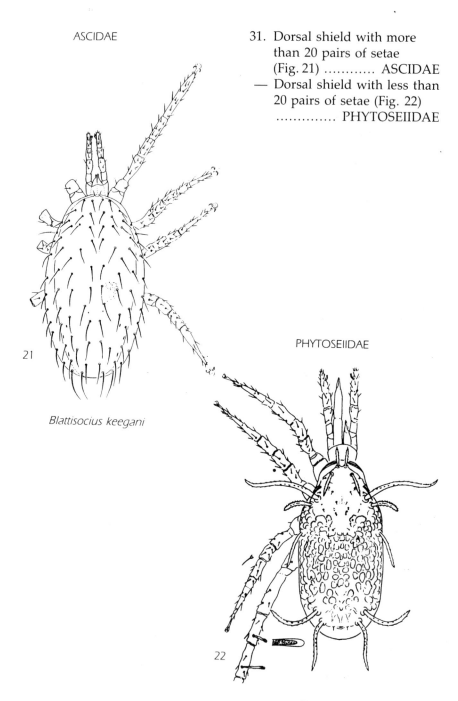

ASCIDAE

21

Blattisocius keegani

PHYTOSEIIDAE

22

Phytoseius ferox

REFERENCES

Atyeo, W.T. (1960) A revision of the mite family Bdellidae in North and Central America (Acarina, Prostigmata). *Univ. Kansas Sci. Bull.*, **40**, 349–499.

Baker, E.W. and Balock, J.W. (1944) Mites of the family Bdellidae. *Proc. Entomol. Soc. Wash.*, **46**, 176–84.

Baker, E.W. and Wharton, G.W. (1952) *An Introduction to Acarology*. Macmillan, New York, pp. 1–465.

Boshko, G.V. and Skylar, V.E. (1981) Parasitization of ixodid ticks on horse flies. *Meditsin. Parazitol.*, **50**, 80–1 (in Russian).

Cook, D.R. (1974) Water mite genera and subgenera. *Mem. Am. Entomol. Inst.*, **21**, 1–860.

Evans, G.O., Sheals, J.G. and Macfarlane, D. (1961) *The Terrestrial Acari of the British Isles*. Trustees of the British Museum, pp. 1–219.

Gerson, U. (1966) A redescription of *Eupalopsis maseriensis* (Canestrini and Fanzago) (Acarina: Eupalopsellidae) *Israel J. Zool.*, **15**, 148–54.

Gerson, U. and Schneider, R. (1981) Laboratory and field studies on the mite *Hemisarcoptes coccophagus* Meyer (Astigmata: Hemisarcoptidae), a natural enemy of armored scale insects. *Acarologia*, **22**, 199–208.

Krantz, G.W. (1978) *A Manual of Acarology*, 2nd edn, Oregon State University Book Stores, Corvallis, pp. 1–509.

Lockwood, J.A. (1987) Entomological warfare: history of the use of insects as weapons of war. *Bull. Entomol. Soc. Am.*, **33**, 76–82.

Manson, D.C.M. (1972) A contribution to the study of the genus *Rhizoglyphus* Claparede, 1869 (Acarina: Acaridae). *Acarologia*, **13**, 621–50.

Mullen, G.R. (1974) Acarine parasites of mosquitoes. II Illustrated larval key to the families and genera of mites reportedly parasitic on mosquitoes. *Mosq. News*, **34**, 183–95.

Oliver, J.H. Jr, McKeever, S. and Pound, J.M. (1986) Parasitism of larval *Ixodes* ticks by chigger mites and fed female *Ornithodoros* ticks by *Ornithodoros* males. *J. Parasitol.*, **72**, 811–12.

Smiley, R.L., Mites (Acari) in *Insect and Mite Pests in Food. An Illustrated Key*, Chapter 1 (ed. J.R. Gorham), USDA and FDA Tech. Bull, no. 3. (in press).

Summers, F.M. (1962) The genus *Stigmaeus* (Acarina: Stigmaeidae) *Hilgardia*, **33**, 491–537.

Zakhavtkin, A.A. (1959) *Fauna of U.S.S.R., Arachnoidea*. Vol. VI, no. 1, Tyroglyphoidea (Acari). American Institute of Biological Sciences, pp. 1–573.

3
Acaridae

DIAGNOSIS: Acaridae are whitish, slow-moving mites whose prodorsum is usually covered with a shield-like sclerite. They carry nude or have slightly barbed opisthosomal setae, and have a sejugal furrow and well-developed empodial claws (O Connor, 1982). Many species are associated with various arthropods and commonly occur in stored foods in which they often constitute major pests (Hughes, 1976).

Tyrophagus putrescentiae (Schrank) is a pest of stored grains, cheese and even fungus cultures in the laboratory (Hughes, 1976). Under moderate temperature conditions and near-saturation humidities, it completed a generation in 2–3 weeks when given various fungi or wheat germ. This mite is considered (Brust and House, 1988) to be an important mortality factor of the southern corn rootworm, *Diabrotica undecimpunctata howardi* Barber (Coleoptera: Scarabaeidae), in peanut and corn fields in North Carolina. *T. putrescentiae* was observed to initiate consumption of rootworm eggs, rapidly located these eggs in the soil, and preferred them over fungi, organic debri and dead arthropods. Significantly fewer adult rootworms emerged from pest eggs placed with mites in soil-filled containers than from the control, mite-less containers. Rack and Rilling (1978) found this mite to feed on dead and living adults and eggs of the grape phylloxera *Daktulosphaira vitifolii* (Fitch) (Homoptera: Phylloxeridae) in the latter's leaf galls. When offered only live insects in the laboratory, the mite required about the same period (2–3 weeks) for its development at 23°C and 85% relative humidity, indicating the suitability of this diet. *T. putrescentiae* is sensitive to high saturation deficits and low temperatures, attributes which preclude its use for the biological control of phylloxera in moderate climates (Rack and Rilling, 1978). Walter et al. (1986) successfully reared different *Tyrophagus* (including *putrescentiae*) on various living nematodes. These data suggest that the potential of *Tyrophagus* spp. as a natural enemy of soil pests should be evaluated further.

Several plant-parasitic nematodes, including *Ditylenchus*, *Heterodera* and *Longidorus*, were observed by Sturhan and Hampel (1977) to be consumed by the bulb mite, *Rhizoglyphus echinopus* (Fumouze and Robin). Small nematodes were completely devoured, whilst larger ones were cut into pieces and sucked out; cysts were attacked only after some initial hesitation. Sturhan and Hampel (1977) suggested a role for *R.*

ACARIDAE

echinopus in regulating nematode populations in the soil. The mite is a polyphagus pest of bulbs, corms and tubers (Hughes, 1976), its development being similar to that of *T. putrescentiae*. Like the latter, it is a pest which could at times be used for biological control purposes.

REFERENCES

Brust, G.E. and House, G.J. (1988) A study of *Tyrophagus putrescentiae* (Acari: Acaridae) as a facultative predator of southern corn rootworm eggs. *Exp. Appl. Acarol.*, **4**, 335–44.

Hughes, A.M. (1976) *The Mites of Stored Food and Houses*, 2nd edn, Her Majesty's Stationery Office, 400 pp.

O Connor, B.M. (1982) Astigmata. In *Synopsis and Classification of Living Organisms* (ed. S.B. Parker), McGraw-Hill, New York, pp. 146–69.

Rack, G. and Rilling, G. (1978) Über das Vorkommen der Modermilbe, *Tyrophagus putrescentiae* (Schrank) in Blattgallen der Reblaus, *Dactylosphaera vitifolii* Shimer. *Vitis*, **17**, 54–66.

Sturhan, D. and Hampel, G. (1977) Pflanzenparasitische Nematoden als Beute der Wurzelmilbe *Rhizoglyphus echinopus* (Acarina, Tyroglyphidae). *Anz. Schädl. Pflazensch. Umwelt.*, **50**, 115–18.

Walter, D.E., Hudgens, R.A. and Freckman, D.W. (1986) Consumption of nematodes by fungivorous mites, *Tyrophagus* spp. (Acarina: Astigmata: Acaridae). *Oecologia*, **70**, 357–61.

4
Anystidae

DIAGNOSIS: The Anystidae are large, reddish, soft-bodied mites which carry few dorsal setae and no prodorsal sensillae. They possess a palpal thumb-claw complex, whose tarsus is longer than its claw. They prey on mites and small insects, and are fast runners.

In North America *Anystis agilis* (Banks) feeds on many pests infesting alfalfa (Frazer and Nelson, 1981), apple and citrus orchards (MacPhee and Sanford, 1961; Mostafa *et al.*, 1975) and vineyards (Sorensen *et al.*, 1976). The mite is easily recognizable by its bright-red colour and characteristic figure-of-eight, whirling running pattern (Muma, 1975). Females consumed a daily average of 5.6 nymphs of the grape leafhopper *Erythroneura elegantula* Osborn (Plate 1). When offered spider mites (*Tetranychus urticae* Koch), *A. agilis* consumed an average of 872.6 females/predator throughout its life, a female mite ingesting 22–40 female spider mites/day. Smaller stages of this prey were taken to a lesser degree; when reared only on spider mites, the predators were sluggish and unco-ordinated (Sorensen *et al.*, 1976). Fecundity on leafhoppers was slightly higher than on spider mites (32.9 vs. 29.1 ova/female). Fresh water as well as plant exudates may also be taken. On citrus *A. agilis* preferred the citrus thrips (*Scirtothrips citri* (Moulton)) over the citrus red mite (*Panonychus citri* McGregor) and was fertile only on the former prey (Mostafa *et al.*, 1975). In California the predator raised two annual generations and was considered to be an important natural enemy of the grape leafhopper. In British Columbia the mite fed on all instars and morphs of aphids on alfalfa, and was usually associated with these pests there (Frazer and Nelson, 1981). *Anystis agilis* was also recorded as an important predator of the woolly pine needle aphid, *Schizolachnus pini-radiata* (Davidson), attacking about 53% of the overwintering eggs (Grobler, 1962). The long generation time and cannibalistic habits of *A. agilis* precluded it from consideration as a candidate for mass-rearing in the laboratory. On the other hand, efforts should be made to conserve its populations in agricultural systems. MacPhee and Sanford (1961) found *A. agilis* to be very sensitive to many common pesticides, an exception being the organophosphorus diazinon.

Anystis baccarum (L.) cleaned up spider mite infestations on blackberries and soybeans within 2 and 5–7 days, respectively, with an initial predator:prey ratio of 1:30, and within seven days on blackberries at an initial ratio of 1:50 (Lange *et al.*, 1974a). This species was released on pest-infested plants immediately after having been obtained in large numbers from oak and pine litter around Moscow (Lange *et al.*, 1974b); the predator usually raises two annual generations in that region. *Anystis salicinus* (L.) was introduced from southern France into Australia in 1965 to control the red-legged earth mite, *Halotydeus destructor* (Tucker), and substantially reduced its populations there (Wallace, 1981). In South Africa this predator is an effective predator of *H. destructor* and of the lucerne flea, *Sminthurus viridis* (L.) (Collembola) (Meyer Smith and Ueckermann, 1987). *Anystis* spp. also fed on tick larvae in the laboratory (Holm and Wallace, 1989), suggesting additional pest control possibilities.

REFERENCES

Frazer, B.D. and Nelson, C. (1981) Note on the occurrence of predatory *Anystis* mites (Acari: Anystidae) in SW British Columbia. *J. Entomol. Soc. British Columbia*, **78**, 46.

Grobler, J.H. (1962) The life history and ecology of the woolly pine needle aphid, *Schizolachnus pini-radiata* (Davidson) (Homoptera: Aphididae). *Can. Entomol.*, **94**, 35–45.

Holm, E. and Wallace, M.M.H. (1989) Distribution of some anystid mites (Acari: Anystidae) in Australia and Indonesia and their role as possible predators of the cattle tick, *Boophilus macroplus* (Acari: Ixodidae). *Exp. Appl. Acarol.*, **6**, 77–83.

Lange, A.B., Drozdovskii, E.M. and Bushkovskaya, L.M. (1974a) The anystis mite – an effective predator of small phytophages. *Zaschita Rast.*, **1974**, 26–8 (in Russian).

Lange, A.B., Drozdovskii, E.M. and Bushkovskaya, L.M. (1974b) Collecting and releasing anystis. *Zaschita Rast.*, **1974**, 33–4 (in Russian).

MacPhee, A.W. and Sanford, K.H. (1961) The influence of spray programs on the fauna of apple orchards in Nova Scotia. XII. Second supplement to VII. Effects on beneficial arthropods. *Can. Entomol.*, **93**, 671–3.

Meyer, M.K.P. (Smith) and Ueckermann, E.A. (1987) A taxonomic study of some Anystidae (Acari: Prostigmata). *Entomol. Mem. Dep. Agric. Wat. Supp. Repub. S. Afr.*, **68**, 1–37.

Mostafa, A.R., DeBach, P. and Fisher, T.W. (1975) Anystid mite: citrus thrips predator. *Calif. Agric.*, **29(3)**, 5.

Muma, M.H. (1975) Mites associated with citrus in Florida. *Univ. Fla. Agric. Exp. St. Bull.*, **640A**.

Sorensen, J.T., Kinn, D.N., Doutt, R.L. and Cate, J.R. (1976) Biology of the mite *Anystis agilis* (Acari: Anystidae): a California vineyard predator. *Ann. Entomol. Soc. Am.*, **69**, 905–10.

Wallace, M.M.H. (1981) Tackling the lucerne flea and red-legged earth mite. *J. Agric. W. Austr.*, **22**, 72–4.

5
Arrenuridae

DIAGNOSIS: The bodies of these mites are heavily sclerotized, with closely fitting dorsal and ventral shields present. The ventral side of the palpal tibia bulges to form an uncate pulp. The coxal plates are united with the ventral shield. The genital aperture is located medially, beyond coxae IV. Genital acetabula on wing-like acetabular plates are incorporated into the ventral shield. Males usually possess a posterior extension incorporated into the dorsal shield.

Larvae of *Arrenurus* spp. are the most common mite parasites of mosquitoes, being associated with several mosquito genera (Smith, 1983). The larvae locate host pupae and parasitize the adult mosquitoes as they emerge. When mosquitoes land on water to oviposit, the mites detach and continue their development in the water. Nymphal and adult mites feed on ostracods (Böttger, 1970). Females of various *Arrenurus* species produced about 55–88 eggs/year (Stechmann, 1978).

Parasitism appears to depend on temporal and spatial co-occurrence of the mites and their hosts (mosquitoes as well as some other aquatic insects) (Smith, 1983). During parasitization a single feeding tube, or stylostome, is formed in the host's body at the site of attachment due to a reaction between mite saliva and insect haemolymph. The stylostome remains in the host's body even after the mites drop off; the number of stylostomes in a given mosquito's body may therefore be used to estimate parasite load in its past. Lanciani (1979) found that the average number of stylostomes was always higher in young as compared to older mosquitoes (*Anopheles crucians* Wiedemann), apparently as a consequence of earlier mortality of mosquitoes which had carried more mites. Smith and McIver (1984) estimated that about 42.5% of newly emerged *Coquillettidia perturbans* (Walker) were missing from the host-seeking populations of this mosquito in Ontario, Canada, as a result of parasitism by *Arrenurus danbyensis* Mullen. Furthermore, *Arrenurus*-parasitized mosquitoes produced progressively fewer eggs, in their first gonotrophic cycle, as the number of mites per host increased (Lanciani, 1983). This was attributed to depletion by mites of energy reserves accumulated in the host for reproduction. Reisen and Mullen (1978) showed that *A. madaraszi* Daday caused high mortality in preoviposition females of the genera *Aedes* and *Culex* in Pakistan; male survival was even further reduced.

ARRENURIDAE

Arrenurus spp. drop off their mosquito hosts as the latter arrive at water bodies for their first oviposition. Consequently these mites occur, almost exclusively, on nulliparous mosquitoes. The presence of *Arrenurus* on mosquitoes may, therefore, be used in age-grading studies, useful for investigations into mosquito behaviour (Corbett, 1970; Jalil and Mitchell, 1972).

Since the basic biology of *Arrenurus* spp., including laboratory rearing methods, is known (Stechmann, 1978), a field effort in reducing mosquito populations with these mites appears to be due. Smith and McIver (1984) advocated the use of *A. danbyensis* against European and Asian *Coquillettidia*.

REFERENCES

Böttger, K. (1970) Die Ernährungsweise der Wassermilben (Hydrachnellae, Acari). *Int. Rev. Ges. Hydrobiol.*, **55**, 895–912.

Corbett, P.S. (1970) The use of parasitic water-mites for age-grading female mosquitoes. *Mosq. News*, **30**, 436–8.

Jalil, M. and Mitchell, R. (1972) Parasitism of mosquitoes by water mites. *J. Med. Entomol.*, **9**, 305–11.

Lanciani, C.A. (1979) Water mite-induced mortality in a natural population of the mosquito *Anopheles crucians* (Diptera: Culicidae). *J. Med. Entomol.*, **15**, 529–32.

Lanciani, C.A. (1983) Overview of the effects of water mite parasitism on aquatic insects. In *Biological Control of Pests by Mites* (eds M.A. Hoy, G.L. Cunningham and L. Knutson), University of California Special Publication no. 3304, pp. 86–90.

Reisen, W.K. and Mullen, G.R. (1978) Ecological observations on acarine associates (Acari) of Pakistan mosquitoes (Diptera: Culicidae). *Environ. Entomol.*, **7**, 769–76.

Smith, B.P. (1983) The potential of mites as biological control agents of mosquitoes. In *Biological Control of Pests by Mites* (eds M.A. Hoy, G.L. Cunningham and L. Knutson), University of California Special Publication no. 3304, pp. 79–85.

Smith, B.P. and McIver, S.B. (1984) The impact of *Arrenurus danbyensis* Mullen (Acari: Prostigmata: Arrenuridae) on a population of *Coquillettidia perturbans* (Walker) (Diptera: Culicidae). *Can. J. Zool.*, **62**, 1121–34.

Stechmann, D.-H. (1978) Eiablage, Parasitismus und postparasitische Entwicklung von *Arrenurus*-Arten (Hydrachnellae, Acari). *Z. Parasitenkd.*, **57**, 169–88.

6
Ascidae

DIAGNOSIS: Ascids may be recognized by the following combination of characters: stigma located between legs III–IV, with straight peritremes; without leg grooves; metasternal shields small; male spermadactyl with digits free distally; legs I with ambulacra; female genital shield not flask-shaped and dorsal shield with more than 23 pairs of setae. These are free-living predators commonly found in soils, on plants and in stored products.

The cosmopolitan *Blattisocius tarsalis* (Berlese) completed a generation (egg to egg) in eight days when offered live eggs of the stored product pest moth *Ephestia cautella* (Walker) at 27°C and 73% relative humidity (Haines, 1981). Starving females survived for 11 days under these conditions. Fed, paired-for-life females deposited approximately 32 eggs each. Immatures consumed four moth eggs during their development, males about 65 eggs throughout their lives, and females 100. Mites dispersed by clinging to various moths, whose longevity could be reduced when more than 8–10 mites/moths were present (White and Huffaker, 1969). *B. tarsalis* was the major factor controlling infestations of *E. cautella* in bagged maize in Kenya (Graham, 1970). Haines (1981) concluded that the mite may control this and possibly other stored-product moth pests (which are the mite's preferred hosts) in the absence of pesticide applications. Such control could especially be effective in the tropics, where the low level of contamination left by predators would only be of minor importance.

Binns (1973) showed that *Arctoseius cetratus* (Sellnick) reduced by about 85% the egg hatch of *Lycoriella auripila* (Winnertz) (Diptera: Sciaridae), a pest of cultivated mushrooms, in small pot experiments, and cited unpublished work showing that the mite may limit populations of another mushroom pest, *Tarsonemus myceliophagus* Hussey. *Platyseius* and *Cheiroseius* spp. feed on mosquito eggs and larvae (Smith, 1983), but no numerical data are available.

Lasioseius parberlesi Tseng is a natural enemy of the major mite pest of rice in Taiwan, *Steneotarsonemus spinki* Smiley (Tseng, 1984). The predator displayed a numerical response to increases in pest populations, and was believed by Tseng (1984) to hold *S. spinki* numbers at an endemic, non-injurious level in southern Taiwan.

Lasioseius scapulatus Kennett completed a generation in six days at 24°C while feeding on the nematode *Aphelenchus avenae* Bastian (Imbriani and Mankau, 1983). Nematode numbers in the rearing cultures decreased as mite numbers increased, and the mite consumed most nematode species offered, except the largest ones. The mite's short life cycle, great appetite, ability to subsist on alternative food (fungi), as well as its density dependence on the prey, were all considered to contribute to the efficacy of *L. scapulatus*. On the other hand, its non-specificity could detract from feeding on pest nematodes. Furthermore, plant-feeding nematodes inhabit the entire root zone, but the mite prefers the uppermost soil strata. Imbriani and Mankau (1983) thus concluded that *L. scapulatus* would be best suited to contribute to nematode control in specialized situations, such as greenhouses. Other ascids are also voracious predators on nematodes (Karg, 1983; Sharma, 1971; Walter, 1987), but their potential role as control agents has not been evaluated under field conditions.

REFERENCES

Binns, E.S. (1973) Predatory mites – neglected allies? *Mushroom J.*, **12**, 540–4.

Graham, W.M. (1970) Warehouse ecology studies of bagged maize in Kenya – II. Ecological observations of an infestation by *Ephestia (Cadra) cautella* (Walker) (Lepidoptera, Phycitidae). *J. Stored Prod. Res.*, **6**, 157–67.

Haines, C.P. (1981) Laboratory studies on the role of an egg predator, *Blattisocius tarsalis* (Berlese) (Acari: Ascidae), in relation to the natural control of *Ephestia cautella* (Walker) (Lepidoptera: Pyralidae) in warehouses. *Bull. Entomol. Res.*, **71**, 555–74.

Imbriani, J.L. and Mankau, R. (1983) Studies on *Lasioseius scapulatus*, a mesostigmatid mite predacious on nematodes. *J. Nematol.*, **15**, 523–8.

Karg, W. (1983) Verbreitung und Bedeutung von Raubmilben der Cohors Gamasina als Antagonisten von Nematoden. *Pedobiologia*, **25**, 419–32.

Sharma, R.D. (1971) Studies on the plant parasitic nematode *Tylenchorhynchus dubius*. *Meded. Landbouww, Wageningen*, **71**, 1–154.

Smith, P.B. (1983) The potential of mites as biological control agents of mosquitoes. In *Biological Control of Pests by Mites* (eds M.A. Hoy, G.L. Cunningham and L. Knutson), University of California Special Publication no. 3304, pp. 79–85.

Tseng, Y.-H. (1984) Mites associated with weeds, paddy rice and upland rice fields in Taiwan. In *Acarology VI* (eds D.A. Griffiths and C.E. Bowman), Ellis Horwood, Chichester, Vol. 2, pp. 770–80.

Walter, D.E. (1987) Life history, trophic behavior, and description of *Gamasellodes vermivorax* n. sp. (Mesostigmata: Ascidae), a predator of nematodes and arthropods in semiarid grassland soils. *Can. J. Zool.*, **65**, 1689–95.

White, E.G. and Huffaker, C.B. (1969) Regulatory processes and population cyclicity in laboratory populations of *Anagasta kuhniella* (Zeller) (Lepidoptera: Phycitidae). I. Competition for food and predation. *Res. Popul. Ecol.*, **11**, 57–83.

7
Bdellidae

DIAGNOSIS: The mouthparts of these mites are snout-like, carrying long palpi which terminate with strong setae and lack the palpal thumb-claw complex. Their prodorsum has two prominent sensilla. They are fairly large (up to 4 mm), red–brown or greenish mites which actively hunt their prey, sometimes tethering them with silken lines (Sorensen *et al.*, 1983).

Bdella depressa Ewing is a voracious feeder on various spider mites (especially the clover mite, *Bryobia praetiosa* Koch) and springtails (Collembola). Development of *B. depressa* from larva to adult in the laboratory (at 90% relative humidity) required 21–30 days at 15°C and 14–21 days at 21°C (Snetsinger, 1956). The predator's larvae consumed at least three eggs or two prey (spider mite) larvae, and each nymph required at least 2–3 prey nymphs or adults to complete its development. Eggs, deposited in dry bark crevices, are the main overwintering stage.

Bdella longicornis L. fed on vine spider mites (at the rate of 1.8–3.3/day) in California early in the spring, before phytoseiid predators were sufficiently abundant to control these pests (Sorensen *et al.*, 1983). This early season reduction in spider mite numbers facilitates their subsequent control by the phytoseiids.

Bdellodes lapidaria Kramer has been known as an efficient predator of the lucerne flea (the collembolan *Sminthurus viridis* (L.)), an important pest of pastures in Australia and South Africa, for more than 50 years (Womersley, 1933). Wallace (1954) eliminated *B. lapidaria* from pastures by applying DDT (which has only a limited effect on the pest) to pastures and reported subsequent lucerne flea population increases of 5- and 19-fold. The presence of more than 20 mites/square metre early in the winter prevented pest outbreaks later in the season (Wallace, 1967). This success prompted the introduction of *B. lapidaria* into South Africa, where it became established and even controlled another pestiferous collembolan, namely *Bourletiella arvalis* Fitch (Wallace and Walters, 1974). The predator does not, however, occupy all areas colonized by the pest in Australia, and another bdellid, *Neomolgus capillatus* (Kramer), was subsequently introduced (Wallace, 1974).

Undetermined bdellids were considered among the major predators of lucerne aphids in Australia (Milne and Bishop, 1987). Muma (1975)

Plate 1. Anystidae. *Anystis agilis* devouring second instar nymph of grape leafhopper. (Courtesy of California Agriculture, Agricultural Experiment Station, University of California, Berkeley.)

Plate 2. Eriophyidae. (a) Flower bud galls of skeletonweed infested in the field by *Aceria chondrillae*, (b) and (c) close up of galls. (Courtesy of E.W. Baker, *An Illustrated Guide to Plant Abnormalities Caused by Eriophyd Mites in North America.*)

Plate 3. Galumnidae. Close up of *Orthogalumna terebrantis* excavating tunnels (arrowed) in waterhyacinth leaves.

Plate 4. Galumnidae. E.W. Baker collecting *Orthogalumna terebrantis* from waterhyacinth infested pond at Shell Experiment Station, Maracay, Venezuela in 1977.

Plate 5. Erythraeidae. A grasshopper, *Condracris rosea*, collected in Taiwan, showing mites *Charletonia taiwanensis* on the posterior wings. (Courtesy of Y.S. Chow, Institute of Zoology, Academia Sinica, Taipei, Taiwan Republic of China.)

Plate 6. Phytoseiidae. Phytoseiid mite predators were released in the orchard on the right but not in defoliated trees. (Courtesy of California Agriculture, Agricultural Experiment Station, University of California, Berkeley.)

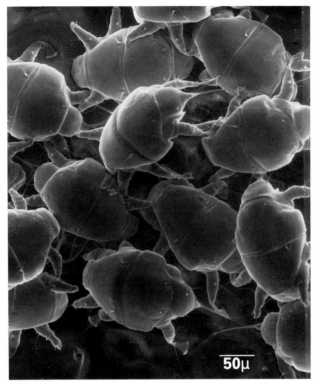

Plate 7. Podapolipidae. Larviform females, *Coccipolipus epilachnae* (scanning electron micrograph).

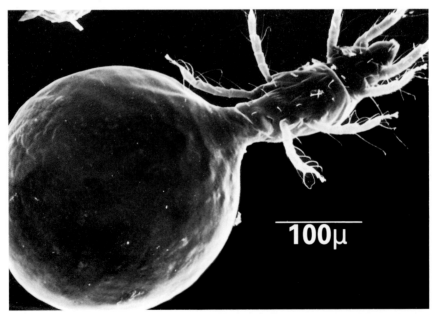

Plate 8. Pyemotidae. A gravid (physogastric) female of *Pyemotes barbara* (scanning electron micrograph).

reported that *Bdella distincta* (Baker and Balock) fed on eggs and crawlers of armoured scale insects in Florida, but no quantitative data were given.

REFERENCES

Milne, W.M. and Bishop, A.L. (1987) The role of predators and parasites in the natural regulation of lucerne aphids in eastern Australia. *J. Appl. Ecol.*, **24**, 893–905.

Muma, M.H. (1975) Mites associated with citrus in Florida. *Univ. Florida Agri. Exp. St., Bull.* 640A.

Snetsinger, R. (1956) Biology of *Bdella depressa*, a predaceous mite. *J. Econ. Entomol.*, **49**, 745–6.

Sorensen, J.T., Kinn, D.N. and Doutt R.L. (1983) Biological observations on *Bdella longicornis*: A predatory mite in California vineyards (Acari: Bdellidae). *Entomography*, **2**, 297–305.

Wallace, M.M.H. (1954) The effect of DDT and BHC on the population of the lucerne flea, *Sminthurus viridis* (L.) (Collembola), and its control by predatory mites, *Biscirus* spp. (Bdellidae). *Aust. J. Agric. Res.*, **5**, 148–55.

Wallace, M.M.H. (1967) The ecology of *Sminthurus viridis* (L.) (Collembola) 1. Processes influencing numbers in pastures in Western Australia. *Aust. J. Zool.*, **15**, 1173–205.

Wallace, M.M.H. (1974) An attempt to extend the biological control of *Sminthurus viridis* (Collembola) to new areas in Australia by introducing a predatory mite, *Neomolgus capillatus* (Bdellidae). *Aust. J. Zool.*, **22**, 519–29.

Wallace, M.M.H. and Walters, M.C. (1974) The introduction of *Bdellodes lapidaria* (Acari: Bdellidae) from Australia into South Africa for the biological control of *Sminthurus viridis* (Collembola). *Aust. J. Zool.*, **22**, 505–17.

Womersley, H. (1933) A possible biological control of the clover springtail or lucerne flea (*Sminthurus viridis* L.) in Western Australia. *J. Aust. Counc. Sci. Indust. Res.*, **6**, 83–91.

8
Camerobiidae

DIAGNOSIS: The Camerobiidae is a small family whose members are identifiable by their long ('stilted') legs and by weak palpi in a ventrally directed gnathosoma, which also carries the looped peritremes. Species of the genus *Neophyllobius* are known to feed on first-instar nymphs ('crawlers') of armoured scale insects (Homoptera: Diaspididae) and on various plant-inhabiting mites.

McGregor (1950) wrote that Pence observed a *Neophyllobius* attacking crawlers, injecting them with an 'opiate' and sucking them dry. Meyer (1962) reported that nymphs and adults of *N. ambulans* Meyer fed on crawlers of the California red scale, *Aonidiella aurantii* (Maskell), infesting orange trees in South Africa. As the predator was rather scarce, it was not considered to be of economic importance in regard to *A. aurantii*.

The possibility that an unnamed *Neophyllobius* plays an important role in the natural control of the European fruit scale, *Quadraspidiotus ostreaeformis* (Curtis), in New Zealand was postulated by Richards (1962). Mites were very abundant wherever the scale was common, and laboratory studies confirmed that *Neophyllobius* was feeding on the crawlers. The latter were initially paralysed by the mites and then sucked dry. Other *Neophyllobius* spp. prey on eriophyoid or tenuipalpid mites (De Leon, 1967; Bolland, 1983), but their effect on the populations of these pests is not known.

Bolland (1986) is revising this family, and the many new species named should facilitate further research into the biocontrol potential of the Camerobiidae.

REFERENCES

Bolland, H.R. (1983) A description of *Neophyllobius aesculi* and its developmental stages (Acari: Camerobiidae). *Entomol. Ber.*, **43**, 42–7.

Bolland, H.R. (1986) Review of the systematics of the family Camerobiidae (Acari: Raphignathoidea). 1 The genera *Camerobia, Decaphyllobius, Tillandsobius* and *Tycherobius*. *Tijd. Entomol.*, **129**, 191–215.

De Leon, D. (1967) *Some Mites of the Caribbean Area*. Allen Press, Lawrence, Kansas, 66 pp.

CAMEROBIIDAE

McGregor, E.A. (1950) Mites of the genus *Neophyllobius*. *Bull. Southern Calif. Acad. Sci.*, **49**, 55–70.

Meyer, M.K.P. (1962) Two new mite predators of the red scale (*Aonidiella aurantii*) in South Africa. *South Afr. J. Agric. Sci.*, **5**, 411–17.

Richards, A.M. (1962) The oyster-shell scale *Quadraspidiotus ostreaeformis* (Curtis), in the Christchurch district of New Zealand. *NZ J. Agric. Res.*, **5**, 95–100.

9
Cheyletidae

DIAGNOSIS: Cheyletid mites have a prominent thumb-claw complex, the palp tarsus carries sickle- and/or comb-like setae, and they lack prodorsal sensilla. They are slow-moving, white, yellow or orange-coloured mites, some of which are ectoparasites of birds, mammals or insects. More often they are free-living predators on other mites, first-instar armoured scale insects (crawlers) or other small insects.

Cheyletus eruditus (Schrank) commonly occurs in stored foods, where it feeds on pest mites (especially *Acarus* spp., suborder Astigmata) and reduces their populations (Norris, 1958; Pulpan and Verner, 1965). The predator, which completes a generation in about four weeks, is currently being used for acarid control in Czechoslovakia (Žďárková, 1986). The combined effects of temperatures and humidities on *C. eruditus* were studied by Solomon (1969), who found that the predator preferred slightly higher temperatures and lower humidities than the prey mite, and that fastest control may occur at 20–25°C and 70–80% relative humidity. The predator survived long periods of food scarcity. It was unresponsive to changes in adult *Acarus* density, but killed more immature prey (Berreen, 1984). Predator augmentation (at *Cheyletus*: *Acarus* ratios of 1:100 to 1:1000) was suggested by Pulpan and Verner (1965) and by Žďárková and Pulpan (1973). The former sieved out *C. eruditus* from predator-rich substrates and released them on grains where predators were rare. The latter authors showed that *C. eruditus* could be kept alive for several months in cold storage and then used successfully. Several acaricides had only a transitory effect on this mite, which has a natural tolerance to certain organophosphate insecticides (Žďárková and Horak, 1987), thereby facilitating integrated stored product mite management. *Cheyletus* appeared to have a limiting effect on populations of the house dust mite (*Dermatophagoides pteronyssinus* (Trouessart)). Repeated vacuum cleaning of a domestic carpet caused an increase in house dust mites, probably by reducing predator numbers (Wassenaar, 1988).

Two separate assessments were made of the ability of *Hemicheyletia bakeri* (Ehara) to control spider mites. Kanavel and Selhime (1967) reported that at 24–30°C development required 22–25 days, between seven and eight prey were daily consumed, and oviposition reached 65 eggs/female. According to Laing (1973), who conducted his experiments at 22°C, fecundity was only 24 eggs/female and spider mite consumption numbers were lower. Weekly countings of mites on strawberry leaves indicated that the numerical response of *H. bakeri* to spider mite population fluctuations was rather weak, and that the predator foraged only within a limited area of the total distribution of the prey.

Cheletogenes ornatus (Canestrini and Fanzago) feeds on armoured scale insect crawlers in many parts of the world. In the laboratory it produced a generation in three months at 28°C, had about a dozen progeny and consumed about 90 crawlers during adult life (Avidov *et al.*, 1968). Mites were sensitive to low humidities and to some insecticides, but, like other cheyletids, they survived periods of starvation. On citrus trees in Israel the populations of *C. ornatus* increased during summer and peaked in autumn.

Neither *Hemicheyletia* nor *Cheletogenes* (nor other cheyletids assayed in the field) were deemed especially promising, but they could be useful as 'second line' predators.

REFERENCES

Avidov, Z., Blumberg, D. and Gerson, U. (1968) *Cheletogenes ornatus* (Acarina: Cheyletidae), a predator of the chaff scale on citrus in Israel. *Isr. J. Entomol.*, **3**, 77–94.

Berreen, J.M. (1984) The functional response of *Cheyletus eruditus* Schrank to changes in the density of its prey, *Acarus siro* L. In *Acarology – VI* (eds D.A. Griffiths and C.E. Bowman), Ellis Horwood, Chichester, Vol. 2, pp. 980–6.

Kanavel, R.F. and Selhime, A.G. (1967) Biological studies on *Paracheyletia bakeri* (Acarina: Cheyletidae). *Fla. Entomol.*, **50**, 107–13.

Laing, J.E. (1973) Evaluating the effectiveness of *Paracheyletia bakeri* (Acarina: Cheyletidae) as a predator of the two-spotted spider mite *Tetranychus urticae*. *Ann. Entomol. Soc. Am.*, **66**, 641–6.

Norris, J.D. (1958) Observations on the control of mite infestations in stored wheat by *Cheyletus* spp. (Acarina: Cheyletidae). *Ann. Appl. Biol.*, **46**, 411–22.

Pulpan, J. and Verner, P.H. (1965) Control of tyroglyphoid mites in stored grain by the predatory mite *Cheyletus eruditus* (Schrank). *Can. J. Zool.*, **43**, 417–32.

Solomon, M.E. (1969) Experiments on predator–prey interactions of storage mites. *Acarologia*, **11**, 484–503.

Wassenaar, D.P.J. (1988) Effectiveness of vacuum cleaning and wet cleaning in reducing house-dust mites, fungi and mite allergen in a cotton carpet: a case study. *Exp. Appl. Acarol.*, **4**, 53–62.

Žďárková, E. (1986) Mass rearing of the predator *Cheyletus eruditus* (Schrank) (Acarina: Cheyletidae) for biological control of acarid mites infesting stored products. *Crop Prot.*, **5**, 122–4.

Žďárková, E. and Horak, E. (1987) Contact acaricides may not restrain effectiveness of the biological control against stored food mites. *Acta Entomol. Bohemoslov.*, **84**, 414–21.

Žďárková, E. and Pulpan, J. (1973) Low temperature storage of the predatory mite *Cheyletus eruditus* (Schrank) for future use in biological control. *J. Stored Prod. Res.*, **9**, 217–20.

10
Cunaxidae

DIAGNOSIS: The cunaxids have 3–5 segmented palpi that are ornamented with strong spines, spurs or apophyses, and which terminate (except in one genus) in a strong claw. The prodorsum carries two pairs of sensory setae (trichobothridia) and tibiae IV have one. These yellow, red or brown mites are fast runners and appear to be indiscriminate feeders on small arthropods which occur on diverse crops as well as in many other habitats. Some cunaxids hunt and fasten their prey with silken threads excreted through their mouth parts (Alberti and Ehrnsberger, 1977).

Cunaxa capreolus (Berlese) was reared by Zaher *et al.*, (1975) on booklice (Psocoptera) and on the oriental spider mite, *Eutetranychus orientalis* (Klein). At 30°C the mite completed a generation in about 4 weeks on both diets, and each female deposited about 45 eggs. Average prey consumption during development was 230 booklice or 472 spider mites/ cunaxid. The predator fed neither on prey eggs nor on several plant diets. Ewing and Webster (1912) considered *Cunaxoides parvus* (Ewing) to be an important predator of the oystershell scale, *Lepidosaphes ulmi* (L.), on apples in North America and *C. oliveri* (Schruft) was observed to feed on the eriophyid gall mite, *Calepitrimerus vitis* (Canestrini), on grape vines in Germany (Schruft, 1971).

REFERENCES

Alberti, G. and Ehrnsberger, R. (1977) Rasterelektronenmikroskopische Untersuchungen zum Spinnvermögen der Bdelliden und Cunaxiden (Acari, Prostigmata). *Acarologia*, **19**, 55–61.

Ewing, H.E. and Webster, R.L. (1912) Mites associated with oyster-shell scale (*Lepidosaphes ulmi* Linne). *Psyche*, **19**, 121–34.

Schruft, G. (1971) *Haleupalus oliveri* nov. spec., eine Dornpalpenmilbe an Reben (*Vitis* spec.) (Acari: Cunaxidae). *Dtsch. Entomol. Z.*, **18**, 377–82.

Zaher, M.A., Soliman, Z.R. and El-Bishlawy, S.M. (1975) Feeding habits of the predaceous mite, *Cunaxa capreolus* (Acarina: Cunaxidae). *Entomophaga*, **20**, 209–12.

11
Eriophyidae

DIAGNOSIS: Eriophyids are minute mites with elongate, annulate, worm-like bodies, two pairs of anteriorly placed legs and a transverse genital aperture located posteriorly to legs II. The annulations may have microtubercles. The dorsal propodosoma is shield-like, with specific patterns. The tarsi are clawless but have a feather-shaped claw-like empodium. The chelicerae are styletiform. These plant pests are known as gall, bud, blister or rust mites, depending on their locality and the specific damage they cause the host plant.

Aceria chondrillae (G. Canestrini) was assayed for control of skeleton weed, *Chondrilla juncea* L., (Plate 2a) in Australia (Carèsche and Wapshere, 1974). Mite feeding induced gall formation in the host's vegetative and flower buds (Plates 2b–c), causing plant stunting, reductions in seed formation and generalized weakness. Generation time was about ten days in summer, and populations reached several hundred/gall. *A. chondrillae* was very tolerant of extreme climates whilst within the galls, withstanding below-freezing as well as 30–35°C temperatures, but humidities above 70% relative humidity were required to establish new infestations. Mites appeared to be specific to certain geographic host forms; a strain originating from Greece was most suitable against skeleton weed in Australia (Carèsche and Wapshere, 1974). This strain did not succeed in the USA; an Italian strain was introduced, which subsequently became established (Sobhian and Andres, 1978).

Several other eriophyids were considered and tried against weeds. These include an *Aceria* sp., which suppressed Russian knapweed, *Centaurea repens* (L.) DC, and was successfully introduced into the Crimea from Central Asia (Kovalev, 1973), *Eriophyes boycei* Keifer, shipped from the USA to Russia for ragweed, *Ambrosia* spp., control (Goeden *et al.*, 1974) and *Aceria convolvuli* (Nalepa), evaluated against field bindweed, *Convolvulus arvensis* L., in the USA (Rosenthal, 1983).

Cromroy (1979) summed up the attributes of eriophyids for the biological control of weeds. These mites are often host-specific, even to the level of plant form and/or host tissue; they are easily disseminated by winds, can be used together with other appropriate biocontrol agents and a few species transmit specific plant viruses. On the other hand, eriophyids are slow-acting and must often be used in concert with another biocontrol agent to bring about weed control. Their sensitivity to low humidities requires special shipping methods (Goeden *et al.*, 1974; Sobhian and Andres, 1978).

Croft and Hoying (1977) noted that apple foliage on which *Aculus schlechtendali* Nalepa, a minor pest of apples, had previously fed, hindered the reproduction of a major pest, the European red mite, *Panonychus ulmi* (Koch). This eriophyid also served as supplementary food for an efficient phytoseiid predator, and was more easily controlled by pesticides. Thus it was suggested that low populations of *A. schlechtendali* should be preserved by selective pesticides.

REFERENCES

Carèsche, L.A. and Wapshere, A.J. (1974) Biology and host specificity of the *Chondrilla* gall mite *Aceria chondrillae* (G. Can.) (Acarina, Eriophyidae). *Bull. Entomol. Res.*, **64**, 183–92.

Croft, B.A. and Hoying, S.A. (1977) Competitive displacement of *Panonychus ulmi* (Acarina: Tetranychidae) by *Aculus schlechtendali* (Acarina: Eriophyidae) in apple orchards. *Can. Entomol.*, **109**, 1025–34.

Cromroy, H.L. (1979) Eriophyoidea in biological control of weeds. In *Recent Advances in Acarology* vol. 1. (ed. J.G. Rodriguez), Academic Press, London, pp. 473–5.

Goeden, R.D., Kovalev, O.V. and Ricker, D.W. (1974) Arthropods exported from California to the U.S.S.R. for ragweed control. *Weed Sci.*, **22**, 156–8.

Kovalev, O.V. (1973) Modern outlooks of biological control of weed plants in the U.S.S.R. and the international phytophagous exchange. In *Proc. 2nd Int. Symp. Biol. Cont. Weeds* (ed. P.H. Dunn), Commonwealth Agricultural Bureaux, pp. 166–72.

Rosenthal, S.S. (1983) Current status and potential for biological control of field bindweed, *Convolvulus arvensis*, with *Aceria convolvuli*. In *Biological Control of Pests by Mites* (eds M.A. Hoy, G.L. Cunningham and L. Knutson), University of California Special Publication no. 3304, pp. 57–60.

Sobhian, R. and Andres, L.A. (1978) The response of the skeletonweed gall midge, *Cystiphora schmidti* (Diptera: Cecidomyiidae), and gall mite, *Aceria chondrillae* (Eriophyidae) to North American strains of rush skeletonweed (*Chondrilla juncea*). *Environ. Entomol.*, **7**, 506–8.

12
Erythraeidae

DIAGNOSIS: The Erythraeidae are large, reddish mites with two pairs of prodorsal sensilla, many dorsal setae, and a thumb-claw process with long, straight chelae. The larvae of these mites usually parasitize other arthropods (Plate 5) whereas their nymphs and adults are predators. A list of erythraeids and their host and prey species is available (Welbourn, 1983, Table 5).

Lasioerythraeus johnstoni Welbourn and Young is a larval parasite and a nymphal and adult predator of the tarnished plant bug, *Lygus lineolaris* (Palisot de Beauvois). Young and Welbourn (1987) considered this mite to have biocontrol potential against the bug because (a) the size of the parasitic larva subequals that of its host, engendering bug mortality; (b) parasitization rates are high (up to 50%) and (c) mite nymphs and adults feed on the same prey.

Balaustium putmani Smiley is the most often evaluated predator in this family. It feeds on many arthropods (including moths and scale insects) and their eggs during all its active stages (Childers and Rock, 1981; Welbourn, 1983). Mite females deposited about 175 eggs when offered eggs of the European red mite, *Panonychus ulmi* (Koch). Best hatch occurred at 12.5–17.5°C, and development to adulthood required 5–6 weeks. Predator larvae consumed less than 20 prey eggs, nymphs about five times as much, and adults 250–350 prey eggs/female (Cadogan and Laing, 1977). Putman (1970) found that each female of *B. putmani* devoured 106 prey eggs/day, or 25 female prey/day. Other pestiferous mites were also consumed, although the predator became entangled in spider mite webbing. The predator could also survive, although not develop, on pollen. Pesticides such as sulphur, DDT and parathion were lethal to *B. putmani* (Herne and Putman, 1966), but more recently it was reported that the predator survives in orchards treated by low-volume pesticide applications (Cadogan and Laing, 1977). Whether *B. putmani* is tolerant to some pesticides or has behavioural escape mechanisms has not been determined, but Hagley and Simpson (1983) reported that the predator (which feeds on the eggs and nymphs of the pear psylla, *Psylla pyricola* Foerster), was very sensitive to permethrin and azinphosmethyl. Childers and Rock (1981) advocated further research into the use of this predator in apple pest management projects.

Delucchi *et al.* (1975) injected larvae of the apple tortricid, *Zeiraphera* diniana Guénée (Lepidoptera), with a radionuclide which was passed on from the emergent females to their eggs. All associated arthropods were then collected and assayed for radioactivity; *Balaustium murorum* Hermann was thus shown to be the most important egg predator of this pest.

REFERENCES

Cadogan, B.L. and Laing, J.E. (1977) A technique for rearing the predaceous mite *Balaustium putmani* (Acarina: Erythraeidae), with notes on its biology and life history. *Can. Entomol.*, **109**, 1535–44.

Cadogan, B.L. and Laing, J.E. (1981) A study of *Balaustium putmani* (Acarina: Erythraeidae) in apple orchards in southern Ontario. *Proc. Entomol. Soc. Ontario*, **112**, 13–22.

Childers, C.C. and Rock, G.C. (1981) Observations on the occurrence and feeding habits of *Balaustium putmani* (Acari: Erythraeidae) in North Carolina apple orchards. *Int. J. Acarol.*, **7**, 63–8.

Delucchi, V., Aeschlimann, J.-P. and Graf, E. (1975) The regulating action of egg predators on the populations of *Zeiraphera diniana* Guénée (Lep. Tortricidae). *Mitt. Schw. Entomol. Gesell.*, **48**, 37–45.

Hagley, E.A.C. and Simpson, C.M. (1983) Effects of insecticides on predators of the pear psylla, *Psylla pyricola* (Hemiptera: Psyllidae), in Ontario. *Can. Entomol.*, **115**, 1409–14.

Herne, D.H.C. and Putman, W.L. (1966) Toxicity of some pesticides to predaceous arthropods in Ontario peach orchards. *Can. Entomol.*, **98**, 936–42.

Putman, W.L. (1970) Life history and behavior of *Balaustium putmani* (Acarina: Erythraeidae). *Ann. Entomol. Soc. Am.*, **63**, 76–81.

Welbourn, W.C. (1983) Potential use of trombidioid and erythraeoid mites as biological control agents of insect pests. In *Biological Control of Pests by Mites* (eds M.A. Hoy, G.L. Cunningham and L. Knutson), University of California Special Publication no. 3304, pp. 103–40.

Young, O.P. and Welbourn, W.C. (1987) Biology of *Lasioerythraeus johnstoni* (Acari: Erythraeidae), ectoparasitic and predaceous on the tarnished plant bug, *Lygus lineolaris* (Hemiptera: Miridae), and other arthropods. *Ann. Entomol. Soc. Am.*, **80**, 243–50.

13
Eupalopsellidae

DIAGNOSIS: Eupalopsellids are recognizable by their very long palpi and chelicerae, the rather reduced palpal thumb-claw complex and by the empodia which resemble two pairs of capitate raylets. They are yellow to orange-coloured mites which usually occur on plants and are often associated with armoured scale insects (Homoptera: Diaspididae). A few species have been collected in the soil.

Saniosulus nudus Summers is the only eupalopsellid whose biology has been studied. When offered Florida red scales (*Chrysomphalus aonidum* (L.)) cultured on green lemons, the mites completed a generation in about three weeks at 24°C; in two weeks at 28°C. All active mite stages fed on diaspidid eggs and crawlers (first-instar nymphs); older scales were also attacked but appeared unaffected. Each female produced about 40–50 eggs, regardless of mating (Gerson and Blumberg, 1969). Fecundity, however, appeared to depend on the scale species fed upon (Zaher *et al.*, 1984). Male and female mites survived for more than three months at 27°C, and consumed an average of 125 and 259 eggs, respectively (Zaher *et al.*, 1984). On citrus trees in Israel, the population of *S. nudus* peaked in late summer and then declined; the predator fed mostly on bark-inhabiting scales (Gerson, 1967). In India large numbers of mites fed on the sugar cane pest, *Melanaspis glomerata* (Green). They crawled under the mother scales and attacked the emerging crawlers there (Rao and Sankaran, 1969). *Saniosulus nudus* is a voracious consumer of diaspidid crawlers in insectary cultures, requiring special chemical control measures (Gerson and Blumberg, 1969). This may indicate its potential as a biocontrol agent under conditions of abundant prey and a suitable climate.

REFERENCES

Gerson, U. (1967) The natural enemies of the chaff scale, *Parlatoria pergandii* Comstock, in Israel. *Entomophaga*, **12**, 97–109.

Gerson, U. and Blumberg, D. (1969) Biological notes on the mite *Saniosulus nudus*. *J. Econ. Entomol.*, **62**, 729–30.

Rao, V.P. and Sankaran, T. (1969) The scale insects of sugar cane. In *Pests of Sugar Cane* (eds J.R. Williams, J.R. Metcalfe, R.W. Mungomery and R. Mathes), Elsevier, Amsterdam, pp. 325–42.

EUPALOPSELLIDAE

Zaher, M.A., Soliman, Z.R. and Rakha, M.A. (1984) Biological studies on the predatory mite *Saniosulus nudus* Summers (Raphignathoidea: Eupalopsellidae). In *Acarology VI* (eds D.A. Griffiths and C.E. Bowman), Ellis Horwood, Chichester, vol. 1, pp. 597–600.

14
Galumnidae

DIAGNOSIS: Galumnid mites are characterized by large, movable pteromorphs, which are anteriorly rounded and posteriorly pointed. The lamellae are weak or lacking and the cuticle is smooth with small porose areas and sclerites around the notogastral setae. The genital plates carry six setae on each plate.

The waterhyacinth mite, *Orthogalumna terebrantis* Wallwork, feeds on various plants of the family Pontederiaceae, but mainly on waterhyacinth, *Eichhornia crassipes* (Mart.) Solms-Laubach, a floating weed of waterways and lakes in many warm parts of the world (Plate 4). Female mites lay eggs in the weed's false leaves (pseudolaminae) and all immature stages live and excavate numerous tunnels therein (Del Fosse, 1978) (Plate 3). Mites developed best and tunnelled most at 10–30°C. An average of 21–24 eggs/female/week were deposited within the range of 10–35°C, but at lower temperatures (5–25°C) there was little mite growth (Del Fosse, 1977a). Development from egg to adult required about 10 days in Argentina (Perkins, 1973).

Direct damage to waterhyacinth is caused by mite feeding and tunnelling, and indirect injury is due to opening up the weed to a phytopathogenic fungus and to promoting the oviposition of a specific phytophagous weevil. The former, *Acremonium zonatum* (Saw.) Gams, which is the causative agent of zonate leaf spot disease of waterhyacinth, gains access to *Eichhornia* tissue through the emergence holes of mite females; the latter also disseminate the fungus. The weevil, *Neochetina eichhorniae* Warner, oviposited significantly more eggs in the presence of *O. terebrantis*. This stimulating effect was attributed to the release of a kairomone from mite-wounded waterhyacinth tissues (Del Fosse, 1977b). Coexisting mites and weevils seem to lead to higher population levels of both organisms (Del Fosse, 1978). When used together against the weed, their combined activity brought about greater control than the sum of each of their effects alone. A combined reduction in waterhyacinth density of about 50% within one year was recorded by Del Fosse (1978).

O. terebrantis thus affects the weed by direct feeding, by promoting the activity of another arthropodan biocontrol agent, and by exposing weed tissue to a phytopathogenic fungus. Despite these attributes, no attempts have been made to augment mite populations in the field in the USA. It is possibly underestimated as a biocontrol agent (Charudattan,

1986). The mite was introduced into eastern Africa and became established there; efforts are currently under way to release it in India and Fiji (Bennett, 1984).

Pergalumna spp. thrive on nematodes, individual mites sometimes feeding on three nematodes successively (Rockett, 1980); some galumnids may thus have potential as biocontrol agents of pest nematodes.

REFERENCES

Bennett, F.D. (1984) Biological control of aquatic weeds. In *Proceedings of the International Conference on Water Hyacinth*, Hyderabad, India, (ed. G. Thyagarajan), UN Environmental Programme, Nairobi, pp. 14–40.

Charudattan, R. (1986) Integrated control of waterhyacinth (*Eichhornia crassipes*) with a pathogen, insects and herbicides. *Weed Sci.*, **34** (Suppl. 1), 26–30.

Del Fosse, E.S. (1977a) Temperature optima for development of *Neochetina eichhorniae* and *Orthogalumna terebrantis*. *Fla. Entomol.*, **60**, 109–13.

Del Fosse, E.S. (1977b) Effect of *Orthogalumna terebrantis* (Acari: Galumnidae) on *Neochetina eichhorniae* (Col. Curculionidae) eggs and oviposition. *Entomophaga*, **22**, 359–63.

Del Fosse, E.S. (1978) Interaction between the mottled waterhyacinth weevil, *Neochetina eichhorniae* Warner and the waterhyacinth mite, *Orthogalumna terebrantis* Wallwork. In *Proc. Int. Symp. Biol. Cont. Weeds*, (ed. T.E. Freeman), University of Florida, pp. 93–7.

Perkins, B.D. (1973) Preliminary studies of a strain of the waterhyacinth mite from Argentina. In *Proc. 2nd Int. Symp. Biol. Cont. Weeds* (ed. R.H. Dunn), Commonwealth Agricultural Bureaux.

Rockett, C.L. (1980) Nematode predation by oribatid mites (Acari: Oribatida). *Int. J. Acarol.*, **6**, 219–24.

15
Hemisarcoptidae

DIAGNOSIS: These are whitish, soft-bodied mites with confluent genital and anal apertures located between and beyond coxae IV, and with an elongate, terminal sucker-like ambulacrum on all tarsi. Their hypopodes (facultative deutonymphs) have a pair of ocelli and their posterior legs are devoid of pretarsi (O'Connor, 1982).

All species of the genus *Hemisarcoptes* are apparently non-specific predators or parasites on armoured scale insects (Homoptera: Diaspididae). Female mites feed on adult scales or their eggs and deposit their own eggs on the adult insects. Some immature mites continue to feed there, while others move to additional hosts. Development of *H. coccophagus* Meyer requires about 16 days at 28°C and 28 days at 21°C (Gerson and Schneider, 1981). Should the host scale become moribund during the larval or first nymphal stages of the mite, the protonymphs molt to become hypopodes (deutonymphs), and may then be disseminated by lady beetles, *Chilocorus* spp. (Coleoptera: Coccinellidae) (Gerson and Schneider, 1981). Hypopodes, which lack mouth parts, can live for up to a fortnight. Immature mites which develop on healthy host scales skip the hypopodal stage, molting directly to tritonymphs.

Species of *Hemisarcoptes* contribute to scale insect mortality and may serve as the key mortality factor, particularly in the absence or scarcity of other natural enemies (Gerson *et al.*, in press). *H. malus* (Shimer) is a major mortality factor of the oystershell scale, *Lepidosaphes ulmi* (L.) on apples in eastern Canada, especially during very cold winters (Lord and McPhee, 1953; Samarasinghe and Leroux, 1966). *H. coccophagus* is very effective in controlling the date palm scale, *Parlatoria blanchardi* (Targioni-Tozzetti), in the Sahel region in Africa (Kaufmann, 1977). These two cases also attest to the wide range of climatic tolerances among mites of this genus. Field populations of *H. coccophagus* (monitored on citrus scale insects and on *Chilocorus*) peaked during late summer in Israel, although some mites were also active in winter (Gerson and Schneider, 1981).

Successful introductions of *Hemisarcoptes* spp. into various parts of the world include the transportation of *H. malus* from eastern to western Canada (Turnbull and Chant, 1961), and its introduction as hypopodes (underneath the elytra of *Chilocorus*) into Bermuda (Bedford, 1949). Methods for mass-rearing *Hemisarcoptes* are available (Gerson *et al.*, in press).

HEMISARCOPTIDAE

Hemisarcoptes malus was susceptible to sulphur and winter oil, but DDT, lead arsenate and summer oil had little effect under field conditions in Canada (Lord and MacPhee, 1953). Sellers and Robinson (1950), in whose laboratory cultures of armoured scale insects the mite became a pest, eliminated *Hemisarcoptes* with Neotran®.

Taxonomic uncertainties have hindered more extensive employment of *Hemisarcoptes* spp. in the past. The revision of this genus by O Connor and Houck (unpublished) should facilitate the use of these mites in the control of armoured scale insects.

REFERENCES

Bedford, E.C.G. (1949) Report of the plant pathologist. In *Report of the Department of Agriculture 1949*, Bermuda Board of Agriculture, pp. 11–19.

Gerson, U. and Schneider, R. (1981) Laboratory and field studies on the mite *Hemisarcoptes coccophagus* Meyer (Astigmata: Hemisarcoptidae), a natural enemy of armored scale insects. *Acarologia*, **22**, 199–208.

Gerson, U., O'Connor, B.A. and Houck, M.A. Acari. In *Armored Scale Insects: Biology, Natural Enemies and Control* (ed. D. Rosen), Elsevier, Amsterdam (in press).

Kaufmann, T. (1977) *Hemisarcoptes* sp. and biological control of the date palm scale, *Parlatoria blanchardi* Targioni, in the Sahel region of Niger. *Environ. Entomol.*, **6**, 882–4.

Lord, F.T. and MacPhee, A.W. (1953) The influence of spray programs on the fauna of apple orchards in Nova Scotia. VI. Low temperature and the natural control of the oystershell scale, *Lepidosaphes ulmi* (L.) (Homoptera: Coccidae). *Can. Entomol.*, **85**, 282–91.

O Connor, B.M. (1982) Astigmata. In *Synopsis and Classification of Living Organisms* (ed. S.B. Parker) McGraw-Hill, New York, pp. 146–69.

Samarasinghe, S. and LeRoux, E.J. (1966) The biology and dynamics of the oystershell scale, *Lepidosaphes ulmi* (L.) (Homoptera: Coccidae), on apple in Canada. *Ann. Entomol. Soc. Quebec*, **11**, 206–92.

Sellers, W.F. and Robinson, G.G. (1950) The effect of the miticide Neotran® upon the laboratory production of *Aspidiotus lataniae* Signoret as a coccinellid food. *Can. Entomol.*, **82**, 170–3.

Turnbull, A.L. and Chant, D.A. (1961) The practice and theory of biological control of insects in Canada. *Can J. Zool.*, **39**, 697–753.

16
Hydryphantidae

DIAGNOSIS: These red water mites have a soft, papillate or lined integument. Coxal plates occur in four groups. The palp is chelate and the palp tibia carries a dorsal distal process, i.e. the distal end of tibia extends beyond insertion of tarsus. Lateral eyes are located in distinct capsules and these lie on the soft integument. A median eye occurs only when a frontal plate is present. The genital field consists of two movable plates and three or more pairs of genital suckers.

In North America many hydryphantids develop in temporary ponds, springs, cold springs and seepage areas (Mullen, 1977), habitats which also abound in mosquitoes, whereas African hydryphantids seem to live in permanent waters (Smith, 1983). The larvae of several genera parasitize mosquitoes, whereas the nymphs and adults are free-living predators feeding on mosquito eggs (Mullen, 1975). The life cycle of *Thyas barbigera* Viets required two years in the north-eastern United States, and 45 eggs or more were produced by each female (Mullen, 1977). Hydryphantids also parasitize black flies (Simuliidae), biting midges (Ceratopogonidae) (Smith and Oliver, 1986) and horse flies (Tabanidae) (Smith, 1983), dipterous families of medical and veterinary importance.

Fenley (1966) believed that the flight activity and host seeking of *Aedes ventrovittis* Dyar were curtailed due to infestation by *Panisopsis* sp. (the mite identification is in doubt according to Smith and Oliver, 1986). Parasitized females were restricted to flying only during the optimal evening hours, whereas mite-free mosquitoes flew and fed throughout the day. Lanciani (1979) demonstrated that heavier hydryphantid loads on a water bug linearly increased the instantaneous death rate of the hosts in the laboratory. To what extent this would also apply to mosquitoes (or other economically important hosts) remains to be seen. Furthermore, postlarval feeding by *Thyas* on mosquito eggs probably does not have a serious impact on prey populations (Mullen, 1977).

Some hydryphantid larvae attach to young emerging mosquitoes, thus parasitizing members of both sexes, whereas others attack only females as these arrive to oviposit. Consequently, only members of the latter group can be used for separation between nulliparous or parous mosquitoes (Mullen, 1977).

REFERENCES

Fenley, W.R. (1966) Field observations of *Aedes ventrovittis* Dyar (Diptera: Culicidae) parasitized by *Panisopsis* sp. (Acarina: Thyasidae). *Mosq. News*, **26**, 583–4.

Lanciani, C.A. (1979) The influence of parasitic water mites on the instantaneous death rate of their hosts. *Oecologia*, **44**, 60–2.

Mullen, G.R. (1975) Predation by water mites (Acarina: Hydrachnellae) on the immature stages of mosquitoes. *Mosq. News*, **35**, 168–71.

Mullen, G.R. (1977) Acarine parasites of mosquitoes. IV. Taxonomy, life history and behavior of *Thyas barbigera* and *Thyasides sphagnorum* (Hydrachnellae: Thyasidae). *J. Med. Entomol.*, **13**, 475–85.

Smith, B.P. (1983) The potential of mites as biological control agents of mosquitoes. In *Biological Control of Pests by Mites* (eds M.A. Hoy, G.L. Cunningham and L. Knutson), University of California Special Publication no. 3304, pp. 79–85.

Smith, I.M. and Oliver, D.R. (1986) Review of parasitic associations of larval water mites (Acari: Parasitengona: Hydrachnida) with insect hosts. *Can. Entomol.*, **118**, 407–72.

17
Laelapidae

DIAGNOSIS: Members of this family can be identified by the following combination of characters: dorsal shield with more than 23 pairs of setae and female genital shield flask-shaped; without leg grooves; stigma located between legs III and IV, with straight peritremes; metasternal shields small; male spermadactyl with digits free distally; leg I usually with ambulacrum. Many species of this family are blood-sucking parasites of birds and mammals, but others are nest-inhabiting or free-living predators of small invertebrates.

Chiang (1970) concluded that soil-borne *Androlaelaps* and *Stratiolaelaps* spp. usually accounted for about 20% of the natural control of northern and western corn rootworms (*Diabrotica longicornis* (Say) and *D. virgifera* LeConte, respectively) in Minnesota. Predators and prey had overlapping vertical distributions, and the laelapids fed on rootworm eggs and larvae in the laboratory. In field experiments mite populations increased threefold with manure applications, apparently bringing about 63% corn rootworm control. The life histories of both laelapids were studied together (Mihm and Chiang, 1976). A generation (egg to egg) required about two months at 20°C and 95% relative humidity, fecundity was 34.4 eggs/female, and some adults lived for as long as 8–9 months. Another soil-borne predator, *Hypoaspis aculeifer* Canestrini, brought about considerable reductions in the populations of the plant nematode *Tylenchorhynchus dubius* (Butchli) in pot experiments (Sharma, 1971; van de Bund, 1972). One mite generation required 3–4 weeks under optimal conditions, but fecundity (89 eggs/female on best food) was much influenced by its arthropod or pollen diet (Lobbes and Schotten, 1980). Other *Hypoaspis* spp. have been associated with phytophagous scarabaeid beetles (Costa, 1971), including the coconut rhinoceros beetle, *Oryctes rhinoceros* (L.), an important pest of coconut in the Pacific region. In a laboratory experiment *Hypoaspis* sp. reduced *Oryctes* egg hatch by about 70%. These mites were released on the Tokelau Islands in the Pacific region as a pest control measure (Swan, 1974).

Haemogamasus pontiger Berlese feeds on acarid mites and on wounded or weakly sclerotized insect larvae, and completed a life cycle in about 10 days at 27°C (Barker, 1968). This mite, collected from bird nests in Norway, was suspected by Mehl (1977) to be an important predator of flea (Siphonaptera) larvae, to the extent of affecting the populations of

LAELAPIDAE

these disease vectors. This possibility was explored by Ryba et al. (1987). Individuals of two flea species (*Neopsylla setosa* (Rotschild) and *Citellophilus simplex* (Wagner)) were placed in nests of a small terrestrial mammal, the European suslik, *Citellus citellus* (L.), and laelapids ((*Eulaelaps stabularis* (Koch)), *Haemogamasus nidi* (Michael) and *Androlaelaps fahrenholzi* (Berlese)) were added to some of these nests. Laelapid-containing nests had significantly fewer fleas than nests devoid of mites, although the reason for this effect is not known. This reduction lasted for several seasons. On the other hand, a higher percentage of flea females in mite-inhabited nests were engorged and carried mature eggs, a phenomenon attributed by Ryba et al. (1987) to mite stimulation of flea behaviour.

REFERENCES

Barker, P.S. (1968) Notes on the bionomics of *Haemogamasus pontiger* (Berlese) (Acarina: Mesostigmata), a predator on *Glycyphagus domesticus* (DeGeer). *Manitoba Entomol.*, **2**, 85–7.

Chiang, H.C. (1970) Effects of manure applications and mite predation on corn rootworm populations in Minnesota. *J. Econ. Entomol.*, **63**, 934–6.

Costa, M. (1971) Mites of the genus *Hypoaspis* Canestrini, 1884 s. str. and related forms (Acari: Mesostigmata) associated with beetles. *Bull. Br. Mus. Zool.*, **21**, 67–98.

Lobbes, P. and Schotten, C. (1980) Capacities of increase of the soil mite *Hypoaspis aculeifer* Canestrini (Mesostigmata: Laelapidae). *Z. ang. Entomol.*, **90**, 9–22.

Mehl, R. (1977) *Haemogamasus pontiger* Berlese (Acari, Mesostigmata) in Norway. *Rhizocrinus*, **9**, 1–4.

Mihm, J.A. and Chiang, H.C. (1976) Laboratory studies on the life cycle and reproduction of some soil- and manure-inhabiting predatory mites (Acarina: Laelapidae). *Pedobiologia*, **16**, 353–63.

Ryba, J., Rodl, P., Bartos, L., Daniel, M. and Cerny, V. (1987) Some features of the ecology of fleas inhabiting the nests of the European suslik (*Citellus citellus* (L.)). II. The influence of mesostigmatid mites on fleas. *Folia Parasitol.*, **34**, 61–8.

Sharma, R.D. (1971) Studies on the plant parasitic nematode *Tylenchorhynchus dubius. Meded. Landbouww. Wageningen*, **71**, 1–154.

Swan, D.I. (1974) A review of the work on predators, parasites and pathogens for the control of *Oryctes rhinoceros* (L.) (Coleoptera: Scarabaeidae) in the Pacific Area. *Misc. Pub. Commonwealth Inst. Biol. Control*, no. 7, p. 64.

van de Bund, C.F. (1972) Some observations on predatory action of mites on nematodes. *Zeszyty Problemowe Postepow Nauk Rolniczych*, **129**, 103–10.

18
Limnesiidae

DIAGNOSIS: The Limnesiid integument is weak. The palpal telofemur has a single peg-like or hair-like seta on the ventral side. The lateral eyes are separated from one another and lie singly, either below the integument or incorporated therein. The coxal plates are arranged in four groups of two each, seldom forming a single group; coxal plate IV is large, triangular, with the acetabula on the posterior lateral edge. There are no claws on the fourth legs.

Tyrrellia circularis Koenike, which attacks biting midges (Diptera: Ceratopogonidae), produces more than 60 eggs/female. The eggs hatch in about 10 days at 23°C; and parasitic larval stage lasts 4–9 days under the same conditions (Lanciani, 1978). Survival of laboratory-infested hosts was found to decrease as mite load increased, and unparasitized ceratopogonids lived 3–4 times longer than those carrying between four and six mites/host (Lanciani, 1986). Nymphal and adult mites occurred in the same habitats and fed on the eggs, larvae and pupae of various ceratopogonids; the large *T. circularis* may consume the entire body contents of smaller midges. The effect of such predation should be added to the effect of parasitism in order fully to evaluate the impact of *T. circularis* on ceratopogonid populations (Lanciani, 1986).

Working in New Britain (Papua New Guinea), Laird (1947) found only few mosquito larvae in natural ponds which contained the predaceous adults of *Limnesia jamurensis* Oudemans. In the laboratory the mite consumed 6.6 eggs (or very small larvae)/day of the mosquito *Anopheles farauti* Laveran and 7.7/day of *Culex pullus* Theobald. When eggs of both mosquitoes were offered together, the mite consumed 17.8/day. These data, combined with his field observations, convinced Laird (1947) that *L. jamurensis* 'plays a useful role' as a predator of the early aquatic stages of the mosquitoes. Smith and Oliver (1986) have questioned the accuracy of this identification.

The attachment of limnesiid larvae to their host's thorax may damage host musculature and thus cause a reduction in flight capability (Smith, 1988).

REFERENCES

Laird, M. (1947) Some natural enemies of mosquitoes in the vicinity of Palmalmal, New Britain. *Trans. R. Soc. New Zealand*, **76**, 453–76.

LIMNESIIDAE

Lanciani, C.A. (1978) Parasitism of Ceratopogonidae (Diptera) by the water mite *Tyrrellia circularis*. *Mosq. News*, **38**, 282–4.

Lanciani, C.A. (1986) Reduced survivorship in *Dasyhelea mutabilis* (Diptera: Ceratopogonidae) parasitized by the water mite *Tyrrellia circularis* (Acariformis: Limnesiidae). *J. Parasitol.*, **72**, 613–14.

Smith, B.P. (1988) Host–parasite interaction and impact of larval water mites on insects. *Ann. Rev. Entomol.*, **33**, 487–505.

Smith, I.M. and Oliver, D.R. (1986) Review of parasitic associations of larval water mites (Acari: Parasitengona: Hydrachnida) with insect hosts. *Can. Entomol.*, **118**, 407–72.

19
Macrochelidae

DIAGNOSIS: Macrochelids may be recognized by having their peritremes looped around the stigma, which are located between legs III and IV, and by the lack of ambulacra on legs I. They do not have leg grooves; females carry small metasternal shields and the spermadactyl digits of males are free distally. These are fast-moving, free-living predators common in habitats which are rich in decaying organic material, including manure. Females are often disseminated by various flies and beetles.

Several species of *Macrocheles* feed on immatures of the house fly (*Musca domestica* L.) and phoretic on the adults. At 30°C *M. muscaedomesticae* (Scopoli) completed a generation in 4–5 days and deposited about 160 eggs/female (Cicolani, 1979). The predator inhabits the outermost layers of poultry manure heaps, where house fly eggs are usually deposited. It prefers eggs, but will also feed on first-instar larvae, with feeding rates of up to 20 immature flies/day. When added to house fly-infested dung substrates, the mite reduced pest populations by about 90% (Rodriguez *et al.*, 1970). Axtell (1963) demonstrated that untreated cattle manure containing *Macrocheles* produced 61–67% fewer flies than manure treated with an acaricide, indicating that the mite is an important component in the natural suppression of house fly populations (Axtell and Rutz, 1986). Keilbach (1978) reported the natural control of the lesser house fly, *Fannia canicularis* (L.), in a small animal stall by *M. muscaedomesticae*.

Other macrochelids feed on the eggs and small larvae of this and other fly pests, such as the stable fly, *Stomoxys calcitrans* (L.), the lesser housefly and the face fly, *Musca autumnalis* De Geer (Anderson, 1983). *M. glaber* (Muller) is an effective predator of the Australian bush fly, *Musca vetustissima* Walker (Wallace *et al.*, 1979). Almost total control was obtained in a laboratory experiment when 50 mites were added to a 1000 ml dung pad with 300–400 fly eggs. Predator activity in the field resulted in substantial reductions of bushfly emergence where dung quality was high, and in total pest suppression when dung quality was lowered. The presence of sufficient dung beetle (Scarabaeidae) vectors was considered to be crucial to the success of the mites.

Wallace and Holm (1983) introduced *M. peregrinus* Krantz into Australia from South Africa to supplement control of the bush fly and buffalo fly, *Haematobia irritans exiqua* De Meijere. This predator, which completes a generation in three days at 27°C, was chosen from various mite candidates on the basis of its wide range of dung beetle vectors. The introduced species rapidly dispersed, colonizing about 180 000 km^2 in the area around each of the two release sites. No obvious *Macrocheles*-associated reductions in buffalo fly numbers were later observed (Doube *et al.*, 1986). This apparent failure was attributed to mite behaviour in the dung pad (including limited searching ability) and to insufficient alternative prey. Krantz (1983) postulated that additional macrochelid species, mostly as yet unstudied, could also be effective pest fly predators. Life history and feeding parameters of various macrochelids were supplied by Cicolani (1979) and by Halliday and Holm (1987).

Macrocheles spp. may also carry propagules of entomopathogenic fungi to their hosts (Schabel, 1982).

A major drawback to the further use of macrochelids in poultry houses is their susceptibility to pesticides applied to manure for fly control. Anderson (1983) advocated the selection of insecticide-resistant fly predators which could be incorporated into integrated fly management programmes.

REFERENCES

Anderson, J.R. (1983) Mites as biological control agents of dung-breeding pests: practical considerations and selection for pesticide resistance. In *Biological Control of Pests by Mites* (eds M.A. Hoy, G.L. Cunningham and L. Knutson), University of California Special Publication no. 3304, pp. 99–102.

Axtell, R.C. (1963) Effect of Macrochelidae (Acarina: Mesostigmata) on house fly production from dairy cattle manure. *J. Econ. Entomol.*, **56**, 317–21.

Axtell, R.C. and Rutz, D.A. (1986) Role of parasites and predators as biological fly control agents in poultry production facilities. *Misc. Publ. Entomol. Soc. Am.*, **61**, 88–100.

Cicolani, B. (1979) The intrinsic rate of natural increase in dung macrochelid mites, predators of *Musca domestica* eggs. *Boll. Zool.*, **46**, 171–8.

Doube, B.M., Macqueen, A. and Huxham, K.A. (1986) Aspects of the predatory activity of *Macrocheles peregrinus* (Acari: Macrochelidae) on two species of *Haematobia* fly (Diptera: Muscidae). *Misc. Publ. Entomol. Soc. Am.*, **61**, 132–41.

Halliday, R.B. and Holm, E. (1987) Mites of the family Macrochelidae as predators of two species of dung-breeding pest flies. *Entomophaga*, **32**, 333–8.

Keilbach, R. (1978) Zusammenbruch einer Plage der Kleinen Stubenfliege (*Fannia canicularis*) durch den phoretischen Parasiten *Macrocheles muscaedomesticae* in Kleintietstall. *Angew. Parasitol.*, **19**, 221–3.

Krantz, G.W. (1983) Mites as biological control agents of dung-breeding flies, with special reference to the Macrochelidae. In *Biological Control of Pests by*

Mites (eds M.A. Hoy, G.L. Cunningham and L. Knutson), University of California Special Publication no. 3304, pp. 91–8.

Rodriguez, J.G., Singh, P. and Taylor, B. (1970) Manure mites and their role in fly control. *J. Med. Entomol.*, **7**, 335–41.

Schabel, H.G. (1982) Phoretic mites as carriers of entomopathogenic fungi. *J. Invert. Pathol.*, **39**, 410–12.

Wallace, M.M.H. and Holm, E. (1983) Establishment and dispersal of the introduced predatory mite, *Macrocheles peregrinus* Krantz, in Australia. *J. Aust. Entomol. Soc.*, **22**, 345–8.

Wallace, M.M.H., Tyndale-Biscoe, M. and Holm, E. (1979) The influence of *Macrocheles glaber* on the breeding of the Australian bushfly, *Musca vetustissima* in cow dung. In *Recent Advances in Acarology* (ed. J.G. Rodriguez), vol. 2, pp. 217–22.

20
Parasitidae

DIAGNOSIS: Female parasitids are recognizable by their two large metasternal shields, which are located just in front of the genital shield, and the males by having the movable and fixed digits of their spermadactyls fused distally. They have no leg grooves and the stigma open between legs III and IV, with straight peritremes. Parasitids are common predators in the soil. They are often dispersed during their deutonymphal stage by various insects, especially beetles and flies.

Pergamasus quisquiliarum Canestrini feeds on the garden symphylan, *Scutigerella immaculata* (Newport) (Scutigerellidae), a serious soil-borne pest of many crops in western Oregon. At 20°C the mite completed a generation in about 17 days (whereas the pest required 87 days), and deposited about 33 eggs, usually attached to nearby plant roots. During its life the mite consumed an average of 14.2 symphylans, actively searching for its prey (Berry, 1973). Although *P. quisquiliarum* appears to be a non-specific predator, its close association with plant roots enhances the probability that it will feed on phytophagous pests. This and the predator's rapid rate of increase led Berry (1973) to conclude that *P. quisquiliarum* could be an important factor in regulating the rate of symphylid population increase in the field.

Poecilochirus monospinosus Wise, Hennessey and Axtell feeds on house fly (*Musca domestica* L.) immatures in poultry manure (Wise *et al.*, 1988). The life cycle required about 17 days at 26.6°C, and average fecundity was 89 eggs/female. Deutonymphal mites destroyed five eggs or four first-instar larvae/day; females, 13 eggs or 24 larvae/day. Later fly instars could not be overcome. This predator occurred mainly in late spring and early summer, and was considered to be an important, albeit short-term, factor in suppressing fly populations (Axtell and Rutz, 1986; Wise *et al.*, 1988).

Karg (1961) reported that several species of *Pergamasus* feed on stored-product pest mites (*Tyrophagus* spp.) in the soil. Harris and Usher (1978) postulated, on the basis of their laboratory results, that the feeding rate of *Pergamasus longicornis* Berlese on soil Collembola (a few of which are pests) may affect the population dynamics of their prey. Bowman (1987, and citations therein) studied the feeding behaviour and digestion of this species.

REFERENCES

Axtell, R.C. and Rutz, D.A. (1986) Role of parasites and predators as biological fly control agents in poultry production facilities. *Misc. Publ. Entomol. Soc. Am.*, **61**, 88–100.

Berry, R.E. (1973) Biology of the predaceous mite, *Pergamasus quisquiliarum*, on the garden symphylan, *Scutigerella immaculata*, in the laboratory. *Ann. Entomol. Soc. Am.*, **66**, 1354–6.

Bowman, C.E. (1987) Studies on feeding in the soil predatory mite *Pergamasus longicornis* (Berlese) (Mesostigmata: Parasitidae) using dipteran and micro-arthropod prey. *Exp. Appl. Acarol.*, **3**, 201–6.

Harris, J.R.W. and Usher, M.B. (1978) Laboratory studies of predation by the grassland mite *Pergamasus longicornis* Berlese and their possible implications for the dynamics of populations of Collembola. *Sci. Proc. Dublin R. Soc.*, Ser. A, **6**, 143–53.

Karg, W. (1961) Ökologische untersuchungen von edaphischen Gamasiden (Acarina, Parasitiformes). *Pedobiologia*, **1**, 77–98.

Wise, G.U., Hennessey, M.K. and Axtell, R.C. (1988) A new species of manure-inhabiting mite in the genus *Poecilochirus* (Acari: Mesostigmata: Parasitidae) predacious on house fly eggs and larvae. *Ann. Entomol. Soc. Am.*, **81**, 209–24.

21
Phytoseiidae

DIAGNOSIS: Phytoseiids are characterized by having an entire dorsal shield with 20 (or less) pairs of setae. They lack leg grooves, legs I carry ambulacra, their stigma open between legs III and IV, the metasternal shields are small, and the digits of the male spermadactyls are distally free. These are fast-running mites which live on plants and in the soil, feeding on small arthropods (including spider mites), as well as on other available diets, such as homopteran honeydew, pollen and, rarely, plant juices (McMurtry and Rodriguez, 1987).

The development of many phytoseiids requires 6–7 days when offered suitable prey at 27°C and 60–90% relative humidity (Tanigoshi, 1982), and some of the highly fecund species, such as *Phytoseiulus persimilis* Athias-Henriot, deposit more than 60 eggs/female (McMurtry and Rodriguez, 1987). Long-range dissemination of this and other spider mite-specific predators is facilitated by their movement into suitable positions and orienting themselves there in such a way as to facilitate wind dispersal. Upon reaching pest-infested leaves, they locate their prey by the latter's kairomones (Sabelis and Dicke, 1985).

Phytoseiids are the best-known predators among the Acari and may easily be mass-reared and shipped (Overmeer, 1985). Several species have attained commercial status. *P. persimilis* is currently being reared and sold for the biological control of spider mites (especially *Tetranychus urticae* Koch) infesting greenhouse crops, mostly cucumbers, peppers, strawberries and some ornamentals, in many parts of the world (McMurtry, 1982). Other species, such as *Typhlodromus occidentalis* Nesbitt and *T. pyri* Scheuten, successfully control spider mites which damage various fruit trees (Plate 6), but their efficacy was formerly curtailed by insecticides. The discovery of pesticide resistance in phytoseiids (Croft and Barnes, 1971; Hoyt, 1969) thus opened the way for various integrated spider mite control programmes with resistant predators at their core (McMurtry, 1982). Several projects intended to select pesticide-resistant phytoseiids are currently under way (Hoy, 1985), with more to be expected as additional species are studied with regard to their potential against other mites or even insects. Relevant examples are two *Amblyseius* spp. which controlled the cyclamen mite, *Phytonemus pallidus* (Banks) (Tarsonemidae), on strawberries (Huffaker and Kennett, 1956), *Amblyseius victoriensis* (Womersley) which controls citrus eriophyids in Australia (Schicha, 1987), *Euseius hibisci* (Chant), an effective predator of the citrus thrips, *Scirtothrips citri* (Moulton) (Tanigoshi *et al.*, 1984), *Amblyseius cucumeris* (Oudemans), used in The Netherlands for control of *Thrips tabaci* Lindeman on sweet pepper (De Klerk and Ramakers, 1986) and *A. barkeri* (Hughes) used against the same pest on cucumbers under glass in Denmark (Hansen, 1988). Phytoseiids also prey on various other pests (McMurtry and Rodriguez, 1987) but there is no clear evidence that they affect their populations.

McMurtry (1982) discussed prospects for promoting the use of phytoseiids. Besides the classical approaches, he also emphasized the need for further biosystematic studies, for determining the control efficacy of two or more co-existing phytoseiids and for testing other species, considering the small percentage (approx. 4%) of phytoseiids considered of importance in spider mite control.

REFERENCES

Croft, B.A. and Barnes, M.M. (1971) Comparative studies on four strains of *Typhlodromus occidentalis*. III. Evaluations of releases of insecticide-resistant strains into an apple orchard ecosystem. *J. Econ. Entomol.*, **64**, 845–950.

De Klerk, M.-L. and Ramakers, P.M.J. (1986) Monitoring population densities of the phytoseiid predator *Amblyeius cucumeris* and its prey after large scale introductions to control *Thrips tabaci* on sweet pepper. *Med. Fac. Landbouww. Rijksuniv. Gent*, **51/3a** 1045–8.

Hansen, L.S. (1988) Control of *Thrips tabaci* (Thysanoptera: Thripidae) on glasshouse cucumber using large introductions of predatory mites *Amblyseius barkeri* (Acarina: Phytoseiidae). *Entomophaga*, **33**, 33–42.

Hoy, M.A. (1985) Recent advances in genetics and genetic improvement of the Phytoseiidae. *Annu. Rev. Entomol.*, **30**, 345–70.

Hoyt, S.C. (1969) Integrated chemical control of insects and biological control of mites on apple in Washington. *J. Econ. Entomol.*, **62**, 74–86.

Huffaker, C.B. and Kennett, C.E. (1956) Experimental studies on predation: Predation and cyclamen-mite populations on strawberries in California. *Hilgardia*, **26**, 191–222.

McMurtry, J.A. (1982) The use of phytoseiids for biological control: progress and future prospects. In *Recent Advances in Knowledge of the Phytoseiidae* (ed. M.A. Hoy), Division of Agricultural Science, University of California Special Publication no. 3284, pp. 23–48.

McMurtry, J.A. and Rodriguez, J.G. (1987) Nutritional ecology of phytoseiid mites. In *Nutritional Ecology of Insects, Mites and Spiders* (eds F. Slansky Jr and J.G. Rodriguez), John Wiley, Chichester, pp. 609–44.

Overmeer, W.P.J. (1985) Rearing and handling. In *Spider Mites, Their Biology, Natural Enemies and Control* (eds W. Helle and M.W. Sabelis), Elsevier, Amsterdam, vol. 1B, pp. 161–70.

Sabelis, M.W. and Dicke, M. (1985) Long-range dispersal and searching behaviour. In *Spider Mites, Their Biology, Natural Enemies and Control* (eds W. Helle and M.W. Sabelis), Elsevier, Amsterdam, vol. 1B, pp. 141–60.

Schicha, E. (1987) *Phytoseiidae of Australia and Neighboring Areas*. Indira Publishing House, Oak Park, p. 187.

Tanigoshi, L.K. (1982) Advances in knowledge of the biology of the Phytoseiidae. In *Recent Advances in Knowledge of the Phytoseiidae* (ed. M.A. Hoy), Division of Agricultural Science, University of California, Special Publication no. 3284, pp. 1–22.

Tanigoshi, L.K., Nishio-Wong, J.Y. and Fargerlund, J. (1984) *Euseius hibisci*: Its control of citrus thrips in southern California citrus orchards. In *Acarology VI*, (eds D.A. Griffiths and C.E. Bowman), Ellis Horwood, Chichester, vol. 2, pp. 699–702.

22
Pionidae

DIAGNOSIS: Pionids have a typically weak and smooth integument, which may sometimes be papillate and bear dorsal and ventral plates. The capitulum is separated from the coxae, with or without anchoring processes. Chelicerae are separated medially. The palp tibia has small setae or protuberances on the inner margin, as well as a sclerotized peg on the distal median end. Coxae often occur in four groups but may be fused into two or three groups. The posterior margin of coxae IV usually have projections or apodemes. The legs have swimming hairs and tarsal claws, usually with clawlets.

Pionid larvae parasitize emerging Chironomidae (Diptera), feed for 1–2 days, then drop off the hosts and molt. The nymphs may survive long periods of dryness in a relatively inactive state, returning to full activity upon being wetted (Smith, 1976). This prolonged nymphal period probably preadapted them to exploiting temporary bodies of water (Smith, 1976). Some species of *Piona* inhabit most available water bodies (Smith, 1976), their ranges thus overlapping those of their prey, larvae of aedine mosquitoes. *Piona nodata* (Muller) produces 100–120 eggs/female (Böttger, 1962). Nymphs and female mites are voracious predators. *Piona* nymphs may destroy up to five first-instar mosquito larvae/day (Mullen, 1974), whereas female mites prey on the fourth-stage larvae, several of these being killed every day; males rarely attack mosquito larvae but will feed on previously subdued prey (Smith, 1983). Other quantitative data are lacking. Species of *Piona* seem to prefer certain species of mosquitoes, but unresolved systematic problems hinder further generalizations (Smith, 1983). Other species of *Piona* feed on small aquatic arthropods, such as cladocerans (Böttger, 1970).

REFERENCES

Böttger, K. (1962) Zur Biologie und Ethologie der einheimischen Wassermilben *Arrenurus (Megaluracarus) globator* (Müll. 1776), *Piona nodata nodata* (Müll. 1776) und *Eylais infundibulifera meridionalis* (Thon 1899) (Hydrachnellae, Acari). *Zool. Jb. Syst.*, **89**, 501–84.

Böttger, K. (1970) Die Ernährungsweise der Wassermilben (Hydrachnellae, Acari). *Int. Rev. ges. Hydrobiol.*, **55**, 895–912.

PIONIDAE

Mullen, G.R. (1974) Predation by water mites (Acarina: Hydrachnellae) on the immature stages of mosquitoes. *Mosq. News*, **35**, 168–71.

Smith, I.M. (1976) A study of the systematics of the water mite family Pionidae (Prostigmata: Parasitengona). *Mem. Entomol. Soc. Can.*, **98**, 1–249.

Smith B.P. (1983) The potential of mites as biological control agents of mosquitoes. In *Biological Control of Pests by Mites* (eds M.A. Hoy, G.L. Cunningham and L. Knutson), University of California Special Publication no. 3304, pp. 79–85.

23
Podapolipidae

DIAGNOSIS: The larviform female podapolipids (Plate 7) are small mites with one to three pairs of legs, and variable body shapes (from vermiform to pear-shaped to sac-like). Males usually have three pairs of legs (seldom four); all legs carry claws and pulvilli. The chelicerae are needle-like and the palpi are rudimentary. All members of this family are parasites of insects. They undergo a shortened life cycle. The sexually mature males hatch from eggs and mate with larviform females, which subsequently molt.

Several species of *Locustacarus* parasitize the large tracheae of various short-horned grasshoppers (Acrididae) in the USA, Kenya and New Zealand (Husband, 1974), and there are indications that their feeding may be detrimental to host health. *Podapolipoides grassi* Berlese is a common, apparently widespread ectoparasite of the migratory locust, *Locusta migratoria* (L.) and other grasshoppers. The mites infest third-instar nymphs and subsequent host stages, usually parasitizing the thorax or anterior abdominal segments, often inhabiting the bases of the wings as well as the genitalia (Gauchat, 1972). The parasite spends its entire life cycle on the locust, dispersing during host mating. Wild populations of the Australian plague locust, *Chortoicetes terminifera* Walker, were found to be heavily infested with mites, without discernible effect on their host; similar infestations in laboratory rearings caused a decrease in locust vigour (Gauchat, 1972).

Chrysomelobia labidomerae Eickwort, which in the USA naturally parasitizes only the chrysomelid beetle *Labidomera clivicollis* (Kirby), required about 12.5 days to develop from egg to adult in the laboratory, and a single female lived for 26 days, depositing about 49 eggs during that period. Dispersal occurred only during host mating (Baker and Eickwort, 1975). The beetle did not appear to be damaged by the parasite (Baker and Eickwort, 1975). The mite was later found in Mexico to parasitize another chrysomelid, the Colorado potato beetle, *Leptinotarsa decemlineata* (Say). As *C. labidomerae* does not occur on this pest in the USA, Drummond *et al.* (1985) treated its Mexican form as a new race, suitable for evaluation against the Colorado potato beetle. Subsequent experiments (Drummond, 1988) showed that heavy mite loads (above 20 parasites/*Leptinotarsa*) reduced beetle dispersal.

Coccipolipus epilachnae Smiley, found to attack the Mexican bean beetle,

PODAPOLIPIDAE

Epilachna varivestis Mulsant (Coccinellidae), in El Salvador (Smiley, 1974), was introduced into the USA. Laboratory tests showed that it did not attack any beneficial (predaceous) coccinellids, being restricted to the phytophagous Epilachninae, and that it could survive the USA winter (Schroder, 1981). In the laboratory *C. epilachnae* completed a generation in 16 days at 20°C, in 10 days at 25°C, but the former temperature appeared to be more optimal; fecundity was about 36 eggs/female (Schroder, 1981). Schroder (1982) reported that *C. epilachnae* reduced beetle oviposition by 66% and increased its mortality by 40%; results which were not supported by later studies (Cantwell *et al.*, 1985). Mites, although very abundant on beetles (100–400/host), caused no significant host mortality. Cantwell *et al.* (1985) attributed these results to different beetle handling and crowding methods in the laboratory. Hochmuth *et al.* (1987), who obtained similarly discouraging results, postulated that the parasitic nature of *C. epilachnae* may have decreased during its prolonged (30 generations) laboratory rearing. Studies on other species in the family which parasitize pests are clearly indicated.

REFERENCES

Baker, T.C. and Eickwort, G.C. (1975) Development and bionomics of *Chrysomelobia labidomerae* (Acari: Tarsonemina; Podapolipidae), a parasite of the milkweed leaf beetle (Coleoptera: Chrysomelidae). *Can. Entomol.*, **107**, 627–38.

Cantwell, G.E., Cantelo, W.W. and Cantwell, M.A. (1985) Effect of a parasitic mite, *Coccipolipus epilachnae*, on fecundity, food consumption and longevity of the Mexican bean beetle. *J. Entomol. Sci.*, **20**, 199–203.

Drummond, F.A. (1988) Spatial pattern analysis of *Chrysomelobia labidomerae* Eickwort (Acari: Tarsonemina; Podapolipidae) on Mexican hosts. *Int. J. Acarol.*, **14**, 199–207.

Drummond, F.A., Logan, P.A., Casagrande, R.A. and Gregson, F.A. (1985) Host specificity tests of *Chrysomelobia labidomerae*, a mite parasitic on the Colorado potato beetle. *Int. J. Acarol.*, **11**, 169–72.

Gauchat, C.A. (1972) A note on *Podapolipoides grassi* Berlese (Acarina: Podapolipidae), a parasite of *Chortoicetes terminifera* Walker, the Australian plague locust. *J. Aust. Entomol. Soc.*, **11**, 259.

Hochmuth, R.C., Hellman, J.L., Dively, G. and Schroder, R.F.W. (1987) Effect of the parasitic mite *Coccipolipus epilachnae* (Acari: Podapolipidae) on feeding, fecundity, and longevity of soybean-fed adult Mexican bean beetles (Coleoptera: Coccinellidae) at different temperatures. *J. Econ. Entomol.*, **80**, 612–16.

Husband, R.W. (1974) Lectotype designation for *Locustacarus trachealis* Ewing and a new species of *Locustacarus* (Acarina: Podapolipidae) from New Zealand. *Proc. Entomol. Soc. Wash.*, **76**, 52–9.

Schroder, R.F.W. (1981) Biological control of the Mexican bean beetle, *Epilachna varivestis* Mulsant, in the United States. In *Biological Control in Crop Production* (ed. G.C. Papavizas), Allanhead, Osmun, pp. 351–60.

Schroder, R.F.W. (1982) Effect of infestation with *Coccipolipus epilachnae* Smiley (Acarina: Podapolipidae) on fecundity and longevity of the Mexican bean beetle. *Int. J. Acarol.*, **8**, 81–4.

Smiley, R.L. (1974) A new species of *Coccipolipus* parasitic on the Mexican bean beetle. *J. Wash. Acad. Sci.*, **64**, 298–302.

24
Pterygosomatidae

DIAGNOSIS: These red, small- to medium-sized mites are characterized by their external peritremes, chelate chelicerae and palpi with thumb-claw complex. Their bodies are either flat and oval, twice as wide as long, or else elongate, and they do not have a suture between the propodosoma and hysterosoma. All known species are parasites of lizards, scorpions or various insects, including cockroaches and triatomine bugs.

Pimeliaphilus plumifer Newell and Ryckman parasitizes bloodsucking bugs of the family Triatomidae. In the laboratory (30°C, 15–20% relative humidity) a minimum of 60 days was required to complete a generation. Females deposited an average of 60 eggs and lived about 54 days (males, 78 days) (Anderson, 1968b). This and other *Pimeliaphilus* species often infest cockroaches in laboratory colonies; when present in large numbers (25/insect or more) they may kill their hosts within one hour (Cunliffe, 1952). Mortality could be due to an injected toxin rather than loss of body fluids, as the dead cockroaches were not exsanguinated. Furthermore, egg capsules produced by heavily parasitized cockroaches often failed to hatch (Field *et al.*, 1966). These mites may also serve as vectors for viruses or bacteria which could be pathogenic to the bugs (Newell and Ryckman, 1966).

Pimeliaphilus spp. do not appear to be very host specific, which would enable them to survive on any triatomine bugs within their range (Anderson 1968a). One species was even successfully transferred from a scorpion to a *Triatoma* (Berkenkamp and Landers, 1983). This could enhance the mite's use for the biological control (directly or by transmitting diseases) of blood sucking bugs.

REFERENCES

Anderson, R.C. (1968a) Ecological observations on three species of *Pimeliaphilus* parasites of Triatominae in the United States (Acarina: Pterygosomidae) (Hemiptera: Reduviidae). *J. Med. Entomol.*, **5**, 459–64.

Anderson, R.C. (1968b) The biology of the conenose bug parasite, *Pimeliaphilus plumifer* Newell and Ryckman (Acarina: Pterygosomidae) (Hemiptera: Reduviidae). *J. Med. Entomol.*, **5**, 473–7.

ILLUSTRATED KEY AND FAMILY DESCRIPTIONS

Berkenkamp, S.D. and Landers, E.J. (1983) Observations on the scorpion parasite *Pimeliaphilus joshuae* Newell and Ryckman, 1966 (Acarina: Pterygosomidae). *J. Arizona-Nevada Acad. Sci.*, **18**, 27–31.

Cunliffe, F. (1952) Biology of the cockroach parasite, *Pimeliaphilus podapolipophagus* Tragardh, with a discussion of the genera *Pimeliaphilus* and *Hirstiella* (Acarina, Pterygosomidae). *Proc. Entomol. Soc. Wash.*, **54**, 153–69.

Field, G., Savage, L.B. and Duplessis, R.J. (1966) Note on the cockroach mite, *Pimeliaphilus cunliffei* (Acarina: Pterygosomidae) infesting oriental, German and American cockroaches. *J. Econ. Entomol.*, **59**, 1532.

Newell, I.M. and Ryckman, R.E. (1966) Species of *Pimeliaphilus* (Acari: Pterygosomidae) attacking insects, with particular reference to the species parasitizing Triatominae (Hemiptera: Reduviidae). *Hilgardia*, **37**, 403–36.

25
Pyemotidae

DIAGNOSIS: Pyemotids can be recognized by their milky-white, spindle-shaped or rounded, segmented bodies. The females possess a pair of pseudostigmatic organs. The chelicerae are stylettiform. Leg I has a claw but usually lacks an empodium. Legs II–IV of the female are similar in size, carrying claws and empodia, and trochanter IV is characteristically subtriangular.

These parasitic mites have a very reduced life cycle. Mated females seek out hosts and attach to them, sometimes paralysing the hosts with an injected toxin. As they feed, the females' opisthosoma distends greatly (Plate 8) and the young develop therein, emerging as adult males and females. Some species of *Pyemotes* (= *Pediculoides*) cause dermatitis in humans. This has hindered use of *Pyemotes*, despite reports of successful pest control in the field (Webster, 1910). Biosystematic research (Moser, 1975; Moser *et al.*, 1987) has shown that the genus *Pyemotes* contains two groups of species: one parasitic only on scolytid beetles and venomless; the other venomous and polyphagous. The latter includes species which are very lethal to insects but of reduced dermatological risk to humans.

Tawfik and Awadallah (1970) reported that *Pyemotes herfsi* (Oudemans) parasitized overwintering caterpillars of the pink bollworm, *Pectinophora gossypiella* (Saunders), in Egypt. Mite-induced mortality reached up to 85% in the first-generation larvae of this important cotton pest. Rizk *et al.* (1979) reported that *P. herfsi* located larvae of the stored-food pest beetle *Tribolium confusum* Duv. beneath 7 cm of bran, and that the progeny of a single mite female parasitized all larvae produced by a single beetle female within ten days.

P. tritici (Lagreze-Fossat and Montane), the straw itch mite, produced a generation in about 7 days and about 100 progeny/female, regardless of humidity; at 50–80% relative humidity more than 200 offspring were produced (Bruce, 1985). The mite attacks and kills various stages of coleopterous and lepidopterous stored-product pests. It was considered for control of the red imported fire ant, *Solenopsis invicta* Buren (Bruce and LeCato, 1980), and a *Pyemotes*-containing commercial product was advertised (Anonymous, 1985). A large-scale field experiment conducted in central Texas failed to demonstrate the efficacy of *P. tritici* against the ant, although more than 1.4 million mites/nest were introduced during a period of 30 weeks (Thorvilson *et al.*, 1987). This and other species of *Pyemotes* have been considered for the control of various pests (Bruce, 1985; Moser, 1975; Webster, 1910). Failures in the use of a *Pyemotes* sp. against a cotton pest in Texas during 1902–3 were documented by Hunter and Hinds (1904).

Bruce (1983) listed the advantages of using *P. tritici* as a biological control agent: mites are easy to rear and their populations can be synchronized (and even kept for later application for about 30 days at low temperatures (Bruce, 1985)); they have a high reproductive potential and a short life cycle, females make up 98% of the population and seek out hosts immediately upon emergence. To this should be added the possibility of rearing *P. tritici* on synthetic media (Bruce, 1989).

Adactylidium sp., which parasitized eggs of the laurel thrips, *Gynaikothrips ficorum* (Marchal), a pest of ornamentals, was believed to exert control on the pest in Egypt (Elbadry and Tawfik, 1966). This or a similar species was an abundant natural enemy of all stages of *G. ficorum* in Brazil, apparently contributing to the pest's control (Bennett, 1965).

REFERENCES

Anonymous (1985) Biological control of *Solenopsis invicta*. *Trop. Pest Manag.*, **31**, 238.

Bennett, F.D. (1965) Observations on the natural enemies of *Gynaikothrips ficorum* Marchal in Brazil. *Commonwealth Inst. Biol. Contr. Tech. Bull.* **5**, 117–25.

Bruce, W.A. (1983) Mites as biological control agents of stored product pests. In *Biological Control of Pests by Mites* (eds M.A. Hoy, G.L. Cunningham and L. Knutson), University of California Special Publication no. 3304, pp. 74–8.

Bruce, W.A. (1985) Temperature and humidity: effects on survival and fecundity of *Pyemotes tritici* (Acari: Pyemotidae). *Int. J. Acarol.*, **10**, 135–8.

Bruce, W.A. (1989) Artificial diet for the parasitic mite *Pyemotes tritici* (Acari: Pyemotidae). *Exp. Appl. Acarol.*, **6**, 11–18.

Bruce, W.A. and LeCato, G.L. (1980) *Pyemotes tritici*: a potential new agent for biological control of the red imported fire ant, *Solenopsis invicta* (Acari: Pyemotidae). *Int. J. Acarol.*, **6**, 271–4.

Elbadry, E.L. and Tawfik, M.S.F. (1966) Life cycle of the mite *Adactylidium* sp. (Acarina: Pyemotidae), a predator of thrips eggs in the United Arab Republic. *Ann. Entomol. Soc. Am.*, **59**, 458–61.

Hunter, W.D. and Hinds, W.E. (1904) The Mexican cotton boll weevil. *USDA Div. Entomol.*, Bull. 45, p. 116.

Moser, J.C. (1975) Biosystematics of the straw itch mite with special reference to nomenclature and dermatology. *Trans. R. Entomol. Soc. London*, **127**, 185–91.

Moser, J.C., Smiley, R.L. and Otvos, I.S. (1987) A new *Pyemotes* (Acari: Pyemotidae) reared from the Douglas-fir cone moth. *Int. J. Acarol.*, **13**, 141–7.

Rizk, G.N., ElBadry, E. and Hafez, S.M. (1979) The effectiveness of predacious and parasitic mites in controlling *Tribolium confusum* Duv. *Mesopotamia J. Agric.*, **14**, 167–82.

Tawfik, M.F.S. and Awadallah, K.T. (1970) The biology of *Pyemotes herfsi* Oudemans and its efficiency in the control of the resting larvae of the pink bollworm *Pectinophora gossypiella* Saunders, in U.A.R. *Bull. Soc. Entomol. Egypte*, **54**, 49–71.

Thorvilson, H.G., Phillips, S.A. Jr, Sorensen, A.A. and Trostle, M.R. (1987) The straw itch mite, *Pyemotes tritici* (Acari: Pyemotidae), as a biological control agent of red imported fire ants, *Solenopsis invicta* (Hymenoptera: Formicidae). *Fla. Entomol.*, **70**, 439–44.

Webster, F.M. (1910) A predaceous and supposedly beneficial mite, *Pediculoides*, becomes noxious to man. *Ann. Entomol. Soc. Am.*, **3**, 15–39.

26
Stigmaeidae

DIAGNOSIS: The Stigmaeidae are red to yellow, ovoid or elongate mites whose dorsum is covered by several plates. They have a palpal thumb-claw complex, short, stylet-like chelicerae but no peritremes. They live in the soil and on plants, and are usually predators of mites. A few prey on scale insects or parasitize flies (Swift, 1987), and *Eustigmaeus* (=*Ledermuelleria*) feeds on mosses (Gerson, 1972).

Species of the genera *Agistemus* and *Zetzellia* are common predators of pestiferous eriophyid, tetranychid and tenuipalpid mites (Santos and Laing, 1985; Oomen, 1982). *Agistemus longisetus* Gonzalez fed on the eggs and active stages of phytophagous mites infesting apple trees in New Zealand. Females deposited two to five eggs/day, became very numerous (100/leaf) during summer and remained active throughout the year (Collyer, 1964). In the laboratory *Agistemus exsertus* Gonzalez devoured an average of 60 eggs/day or 45 immatures/day of tomato russet mites, *Aculops lycopersici* (Massee), at 30°C (Osman and Zaki, 1986). Feeding only upon the eriophyid *Aculus schlechtendali* (Nalepa), *Zetzellia mali* (Ewing) produced a generation in 21 days (at 19°C), deposited an average of 1.7 eggs/day/female (intrinsic rate of increase (R_m) was 0.109 females offspring/female/day) (White and Laing, 1977a). Field observations (White and Laing, 1977b) indicated that the predator cannot control large populations of phytophagous mites by itself. However, populations of *Z. mali* tracked those of eriophyids better than other predators in Ontario (Woolhouse and Harmsen, 1984), and it was tolerant to insecticides and fungicides commonly used in southern Ontario apple orchards (but not to acaricides). *Z. mali* may thus provide a significant level of pest mite control under natural conditions.

Oomen (1982) concluded that several stigmaeids can effectively control the tenuipalpid *Brevipalpus phoenicis* (Geijskes) on tea in Indonesia. These predators readily fed on the tenuipalpid, reproduced faster, and suppressed populations of this pest under laboratory as well as field conditions.

REFERENCES

Collyer, E. (1964) Phytophagous mites and their predators in New Zealand orchards. *NZ J. Agric. Res.*, **7**, 551–68.

STIGMAEIDAE

Gerson, U. (1972) Mites of the genus *Ledermuelleria* (Prostigmata: Stigmaeidae) associated with mosses in Canada. *Acarologia*, **13**, 319–43.

Oomen, P.A. (1982) Studies on population dynamics of the scarlet mite, *Brevipalpus phoenicis*, a pest of tea in Indonesia. *Med. Landbouww. Wageningen*, **82-1**, 1–88.

Osman, A.A. and Zaki, A.M. (1986) Studies on the predation efficiency of *Agistemus exsertus* Gonzalez (Acarina, Stigmaeidae) on the eriophyid mite *Aculops lycopersici* (Massee). *Anz. Schädling. Pflanzenschutz. Umwelt*, **59**, 135–6.

Santos, M.A. and Laing, J.E. (1985) Stigmaeid predators. In *Spider Mites, Their Biology, Natural Enemies and Control* (eds W. Helle and M.W. Sabelis), Elsevier Science Publishers, Amsterdam, vol. 1B, pp. 197–203.

Swift, S.F. (1987) A new species of *Stigmaeus* (Acari: Prostigmata: Stigmaeidae) parasitic on phlebotomine flies (Diptera: Psychodidae). *Int. J. Acarol.*, **13**, 239–44.

White, N.D. and Laing, J.E. (1977a) Some aspects of the biology and a laboratory life table of the acarine predator *Zetzellia mali*. *Can. Entomol.*, **109**, 1275–81.

White, N.D. and Laing, J.E. (1977b) Field observations of *Zetzellia mali* (Ewing) (Acarina: Stigmaeidae) in southern Ontario apple orchards. *Proc. Entomol. Soc. Ontario*, **108**, 23–30.

Woolhouse, M.E.J. and Harmsen, R. (1984) The mite complex of the foliage of a pesticide-free apple orchard: population dynamics and habitat associations. *Proc. Entomol. Soc. Ontario*, **115**, 1–11.

27
Tarsonemidae

DIAGNOSIS: Tarsonemids are small mites with broad to elongated oval bodies; the integument is hard and shiny. Females have a clavate pseudostigmatic organ. Legs IV differ from other legs, ending with apical and subapical whip-like setae. Legs IV of males usually terminate with a large single tarsal claw.

Tarsonemids have only two active stages, larvae and adults. Their diets are highly variable. Some feed on green plants (a few are important agricultural pests), others prefer fungi, and several are associates of arthropods, whether as parasites, predators or of some undetermined relationship (Lindquist, 1986).

Iponemus spp. parasitize eggs of pestiferous bark beetles of the genus *Ips* (Scolytidae). They produce 40–80 eggs/female and require about two weeks to complete a generation under summer field conditions (Lindquist, 1969). Various species of *Iponemus* have different effects on beetle populations, and mite-induced egg mortality was reported to be 1–90%. Such mortality, however, could be compensated for by a decrease in larval intraspecific competition and mortality, which in turn detracts from the value of the mites as factors in the pests' biocontrol (Lindquist, 1969). *Iponemus* spp. can survive for extended periods without feeding (Lindquist, 1969), a characteristic which could be of use in the manipulation of these (and other tarsonemid) mites for biological pest control.

Acaronemus destructor (Smiley and Landwehr) feeds on eggs of phytophagous mites (Tenuipalpidae and Tetranychidae) infesting pines in California (Smiley and Landwehr, 1976). Mite numbers were highest in September, coinciding with the population peak of its main tenuipalpid prey, but the actual effect of the predator on its prey is not known.

Fungivorous, non-pestiferous tarsonemids may serve as supplementary diets for some predaceous phytoseiid mites (McMurtry et al., 1970). Certain phytophagous members of this family were recorded from weeds and could, following suitable screening, be considered for the control of these plants. Tarsonemid mites may thus have several roles to play as biological control agents.

TARSONEMIDAE

REFERENCES

Lindquist, E.E. (1969) Review of Holarctic tarsonemid mites (Acarina: Prostigmata) parasitizing eggs of pine bark beetles. *Mem. Entomol. Soc. Can.*, **60**, 1–111.

Lindquist, E.E. (1986) The world genera of Tarsonemidae (Acari: Heterostigmata): a morphological, phylogenetic and systematic revision, with a reclassification of family-group taxa in the Heterostigmata. *Mem. Entomol. Soc. Can.*, **136**, 1–517.

McMurtry, J.A., Huffaker, C.B. and van de Vrie, M. (1970) Ecology of tetranychid mites and their natural enemies. A review. 1. Tetranychid enemies: their biological characters and the impact of spray practices. *Hilgardia*, **40**, 331–90.

Smiley, R.L. and Landwehr, V.R. (1976) A new species of *Tarsonemus* (Acarina: Tarsonemidae), predaceous on tetranychoid mite eggs. *Ann. Entomol. Soc. Am.*, **69**, 1065–72.

28
Tetranychidae

DIAGNOSIS: Spider mites may be recognized by their needle-like chelicerae, by having a thumb-claw complex, two pairs of eyes and usually two sets of duplex setae on tarsus I. Many species produce various amounts of silk (webbing). They are important plant pests, which suggested their use for weed control (Andres, 1983). Cromroy (1983) has provided a partial list of spider mites obtained from weeds.

Tetranychus desertorum Banks was inadvertently introduced into Australia with shipments of insects intended to control prickly pear, *Opuntia* spp. The mite rapidly colonized cactus-inhabited areas in Queensland and its feeding restricted prickly pear fruiting. According to Dodd (in Hill and Stone, 1985), the mite thinned an area of some 360 ha of 1.5 m tall *O. inermis* DC. by about 75% in two years. The subsequent introduction of more-efficient cactus-feeding insects caused a decline in *T. desertorum*.

Cooreman (1959) observed waterhyacinth, *Eichhornia crassipes* (C. Martius) Solms Laubach, to be heavily damaged by a *Tetranychus* sp. in tropical Africa and suggested that the mite be used for the weed's control. Pieterse (1972) assayed twelve species of polyphagous tetranychids for waterhyacinth control and reported that some of them seriously curtailed the growth of the weed. Pieterse (1972) believed that due to their pestiferous nature, non-specific spider mites could be tried for waterhyacinth control only at great distances from agricultural districts. Hill (1983) advocated the use of *T. lintearius* Dufour for the control of gorse (*Ilex europaeus* L.) in New Zealand. One generation of the mite required 46 days at 15°C and 27 days at 25°C. Each female produced about 50 progeny. There was no winter diapause (Stone, 1986). The strongly webbing colonies move slowly along gorse shoots, feeding and ovipositing. The webbing can envelop entire plants, initially causing foliage bronzing; later branches wilt and sometimes the whole plant eventually dies.

Uncertainties about spider mite identity, reproductive isolation and host range have hindered the wide use of these plant pests in weed control (Hill and Stone, 1985). The growing realization that at least some tetranychids may in fact be quite host-specific should promote their future use in weed control (Hill and Stone, 1985).

TETRANYCHIDAE

REFERENCES

Andres, L.A. (1983) Considerations in the use of phytophagous mites for the biological control of weeds. In *Biological Control of Pests by Mites* (eds M.A. Hoy, G.L. Cunningham and L. Knutson), University of California Special Publication no. 3304, pp. 53–6.

Cooreman, J. (1959) *Tetranychus telarius* (Linne) (Acari, Tetranychidae) parasite de la jacinthe d'eau, Eichhornia crassipes Mart. *Bull. Agric. Congo Belge et Ruanda-Urundi*, **50**, 395–402.

Cromroy, H.L. (1983) Potential use of mites in biological control of terrestrial and aquatic weeds. In *Biological Control of Pests by Mites* (eds M.A. Hoy, G.L. Cunningham, and L. Knutson), University of California Special Publication no. 3304, pp. 61–6.

Hill, R.L. (1983) Prospects for the biological control of gorse. *Proc. 36th NZ Weed Pest Conf.*, pp. 56–8.

Hill, R.L. and Stone, C. (1985) Spider mites as control agents for weeds. In *Spider Mites: Their Biology, Natural Enemies and Control* (eds W. Helle and M.W. Sabelis), vol. 1B, pp. 443–8.

Pieterse, A.H. (1972) A preliminary investigation on control of water hyacinth by spider mites. *Proc. 11th British Weed Control Conf.*, **1**, 1–3.

Stone, C. (1986) An investigation into the morphology and biology of *Tetranychus lintearius* Dufour (Acari: Tetranychidae). *Exp. Appl. Acarol.*, **2**, 173–86.

29
Trombidiidae

DIAGNOSIS: These large, usually red mites can be recognized by their dense coat of setae, by short, non-retractable chelicerae and by having only a single pair of trichobothria.

Eutrombidium locustarum (Walsh) is a parasite and predator of many acridid and tettigoniid grasshoppers in North America (Welbourn, 1983). Hatching mite larvae parasitize grasshoppers from their first nymphal instar on, about 35 mites/host having been recorded (Severin, 1944). Mite feeding seems to interfere with normal adult flight by causing grasshopper wings to become battered and even broken. Parasitism in the field progressed from one grasshopper species to another as each suitable host appeared during the season (Huggans and Blickenstaff, 1966). Mite nymphs and adults seek out grasshopper eggs in the soil and devour them; a pair of *E. locustarum* consumed an average of five eggs/day. In the laboratory the mite can produce a generation in about 60 days, but under field conditions in North Dakota it usually has only one complete and one partial annual generation. The mite deposited an average of 4700 eggs/female (Severin, 1944).

Nymphs and adults of the European *Allothrombium monspessulanum* Robaux and Aeschlimann feed on several pests, especially aphids, and each mite may daily kill several prey. A census conducted by Aeschlimann and Vitou (1986) indicated that where mites were abundant, aphids were scarce, and *vice versa*. *Allothrombium pulvinum* Ewing is an important parasite of aphids in China, because its larvae emerge during April before other natural enemies attack these pests (Zhang, 1988). Two or more larvae/aphid kill the host in 1–3 days, while the parasitism of a single mite reduces aphid fecundity by half. Adult *A. pulvinum* may kill 2.5 aphids in an hour. Small-scale releases indicated that a mite:aphid ratio of 3:1 could control these pests (Zhang, 1988). This is a univoltine species; development of the immatures required 74 days in the laboratory at 20–30°C (Zhang and Xin, 1989). These authors proposed that populations of the predator should be conserved by selective chemicals, and advocated its introduction into new localities. Welbourn (1983, Table 7) listed several Trombidiidae which have a high potential for the biological control of insect pests, and require immediate evaluation. One of their important attributes is that they may be mass-reared in the laboratory for field releases (Welbourn, 1983).

TROMBIDIIDAE

REFERENCES

Aeschlimann, J.P. and Vitou, J. (1986) Observations on the association of *Allothrombium* sp. (Acari: Trombidiidae) mites with lucerne aphid populations in the Mediterranean region. In *Ecology of Aphidophaga* (ed. I. Hodek), Academia, Prague, pp. 405–10.

Huggans, J.L. and Blickenstaff, C.C. (1966) Parasites and predators of grasshoppers in Missouri. *Miss. Agric. Exp. St, Res. Bull.*, **903**, 1–40.

Severin, H.C. (1944) The grasshopper mite, *Eutrombidium trigonum* (Hermann) an important enemy of grasshoppers. *South Dakota Agric. Exp. St, Tech. Bull.*, **3**, 1–36.

Welbourn, W.C. (1983) Potential use of trombidioid and erythraeoid mites as biological control agents of insect pests. In *Biological Control of Pests by Mites* (eds M.A. Hoy, G.L. Cunningham and L. Knutson), University of California, Special Publication no. 3304, pp. 103–40.

Zhang, Z. (1988) Progresses and future prospects of Trombidioidea and Erythraeoidea mites as candidates for biological control of insect pests. *Chin. J. Biol. Cont.*, **4**, 79–82 (in Chinese).

Zhang, Z.-Q. and Xin, J.-L. (1989) Biology of *Allothrombium pulvinum* (Acariformes: Trombidiidae), a potential biological control agent of aphids in China. *Exp. Appl. Acarol.*, **6**, 101–9.

30
Tydeidae

DIAGNOSIS: Tydeids are small, soft-bodied mites with needle-like chelicerae whose bases are fused or contiguous. They have no palpal thumb-claw complex. Their idiosoma is striated and sometimes partly reticulated, and they bear a pair of prodorsal sensilla. These fast-moving mites are commonly found in the soil and on plants, where they feed on a diversity of natural foods.

Homeopronematus anconai (Baker) feeds on various stages of the tomato russet mite, *Aculops lycopersici* (Massee), significantly reducing their population densities (Hessein and Perring, 1986). Predator-free tomato seedlings died due to the pest's attack, while those with the tydeids remained healthy. The mite also feeds on pollen, fungi and even plant tissue. Knop and Hoy (1983b) reared *H. anconai* on various pollens at 24°C and 30°C, at 40–75% relative humidity. Females had a mean generation time of 21 days, and deposited about 16 eggs/mite at the lower temperature. At 30°C they produced about 45 eggs/female and the mean generation time was 12 days. The sex ratio was 2F:1M. In the field the mites served as alternate prey for phytoseiid predators of spider mites. During a year *H. anconai* produced about ten overlapping generations in the San Joaquin Valley of California (Knop and Hoy, 1983a). *H. anconai* is sensitive to many acaricides, but Royalty and Perring (1987) found that avermectin B_1 may control the tomato russet mite without reducing tydeid numbers.

Other tydeids were reported to feed on eriophyids (Schruft, 1972), nematodes (Santos *et al.*, 1981) and other invertebrates. Brickhill (1958) offered scale insect and spider mite eggs to two tydeids and reported reasonable development only on the latter food. Hatch of the spider mite eggs which had been fed-upon however, was close to 100%.

Mendel and Gerson (1982) demonstrated that acaricide-mediated preclusion of *Lorryia formosa* Cooreman from a citrus grove infested by the soft scale *Saissetia oleae* (Olivier) resulted in much heavier contamination of sooty-mould there, as compared to groves whence the mite had not been excluded. *L. formosa* is known to feed on scale-produced honeydew, and was postulated by Mendel and Gerson (1982) to act as a sanitizing agent in citrus groves.

Tydeids thus play three separate beneficial roles: they prey on certain pests, may serve as alternate food for other predators and can 'clean up' after honeydew producers, thereby reducing damage attributable to sooty-mould.

REFERENCES

Brickhill, C.D. (1958) Biological studies of two species of tydeid mites from California. *Hilgardia*, **27**, 601–20.

Hessein, N.A. and Perring, T.M. (1986) Feeding habits of the Tydeidae with evidence of *Homeopronematus anconai* (Acari: Tydeidae) predation of *Aculops lycopersici* (Acari: Eriophyidae). *Int. J. Acarol.*, **12**, 215–21.

Knop, N.F. and Hoy, M.A. (1983a) Tydeid mites in vineyards. *Calif. Agric.*, **37**, (11/12), 16–18.

Knop, N.F. and Hoy, M.A. (1983b) Biology of a tydeid mite, *Homeopronematus anconai* (n. comb.) (Acari: Tydeidae), important in San Joaquin Valley vineyards. *Hilgardia*, **51**, 1–30.

Mendel, Z. and Gerson, U. (1982) Is the mite *Lorryia formosa* Cooreman (Prostigmata: Tydeidae) a sanitizing agent in citrus groves? *Acta Oecol.*, **3**, 47–51.

Royalty, R.N. and Perring, T.M. (1987) Comparative toxicity of acaricides to *Aculops lycopersici* and *Homeopronematus anconai* (Acari: Eriophyidae, Tydeidae). *J. Econ. Entomol.*, **80**, 348–351.

Santos, P.F., Phillips, J. and Whitford, W.G. (1981) The role of mites and nematodes in early stages of buried litter decomposition in a desert. *Ecology*, **62**, 664–9.

Schruft, G. (1972) Les tydeides (Acari) sur vigne. *OEPP/EPPO Bull.*, **3**, 51–5.

31

Uropodidae

DIAGNOSIS: The Uropodidae live in forest and other rich organic soils and in manure. They can be recognized by their obvious leg grooves and by the stigma which are positioned between legs II and III. Many species disperse during their deutonymphal stage, when they attach to insects by means of an anal pedicel.

Fuscuropoda vegetans (De Geer) completed a generation in 32 days at 27°C, deposited about seven eggs/day when fed nematodes and four eggs/day on a diet of house fly (*Musca domestica* L.) immatures. The adults lived for 7–8 months (Jalil and Rodriguez, 1970). The mite, which is sensitive to low humidities, is a slow-moving inhabitant of the below-surface, humid zones of manure heaps, the same strata in which newly hatched fly larvae abound (Willis and Axtell, 1968). Manure populations of *F. vegetans* peaked in autumn and in spring, but were scarce during winter (Peck and Anderson, 1969; Rodriguez *et al.*, 1970). Deutonymphs, males and females feed on first-instar house fly larvae. The deutonymphs reduced larval fly populations by 19.1% in a laboratory experiment, the adults by 35.4% (O'Donnell and Axtell, 1965). Rodriguez *et al.* (1970) obtained about 87% immature house fly mortality when 200 mites were placed in a container with 1000 fly eggs. In another experiment 20 mites with 250 fly eggs accounted for 26.8% pest mortality (Willis and Axtell, 1968).

F. vegetans reproduced while feeding on several diets, including bread, yeasts and nematodes, and was most fecund on a diet of nematodes and house fly immatures (Willis and Axtell, 1968). It cannot, however, subdue the larger second and third instar larvae (Axtell and Rutz, 1986). The mite also consumed eggs of the little house fly, *Fannia canicularis* (L.), causing 14.1–17.7% mortality in laboratory populations (O'Donnell and Nelson, 1967). The mite is phoretic on dung beetles (Willis and Axtell, 1968), which facilitate its access to suitable habitats.

Karg (1986) studied and summarized the feeding habits of uropodids, noting that many are carnivorous.

REFERENCES

Axtell, R.C. and Rutz, D.A. (1986) Role of parasites and predators as biological fly control agents in poultry production facilities. *Misc. Publ. Entomol. Soc. Am.*, **61**, 88–100.

Jalil, M. and Rodriguez, J.G. (1970) Biology of and odor perception by *Fuscuropoda vegetans* (Acarina: Uropodidae), a predator of the house fly. *Ann. Entomol. Soc. Am.*, **63**, 935–8.

Karg, W. (1986) Vorkommen und Ernährung der Milbencohors Uropodina (Schildkrötenmilben) sowie ihre Eignung als Indikatoren in Agroökosystemen. *Pedobiologia*, **29**, 285–95.

O'Donnell, A.E. and Axtell, R.C. (1965) Predation by *Fuscuropoda vegetans* (Acarina: Uropodidae) on the house fly (*Musca domestica*). *Ann. Entomol. Soc. Am.*, **58**, 403–4.

O'Donnell, A.E. and Nelson, E.L. (1967) Predation by *Fuscuropoda vegetans* (Acarina: Uropodidae) and *Macrocheles muscaedomesticae* (Acarina: Macrochelidae) on the eggs of the little house fly, *Fannia canicularis*. *J. Kansas Entomol. Soc.*, **40**, 441–3.

Peck, J.H. and Anderson, J.R. (1969) Arthropod predators of immature Diptera developing in poultry droppings in Northern California. Part 1. Determination, seasonal abundance and natural cohabitation with prey. *J. Med. Entomol.*, **6**, 163–7.

Rodriguez, J.G., Singh, P. and Taylor, B. (1970) Manure mites and their role in fly control. *J. Med. Entomol.*, **7**, 335–41.

Willis, R.R. and Axtell, R.C. (1968) Mite predators of the house fly: a comparison of *Fuscuropoda vegetans* and *Macrocheles muscaedomesticae*. *J. Econ. Entomol.*, **61**, 1669–74.

PART TWO
Acari as Biocontrol Agents

In this second part are chapters which supplement the first, together with reviews presented by Hoy *et al.* (1983). While the material included in Part One treated the predaceous mite families individually, in a 'vertical' manner, Part Two is meant to supply a 'horizontal', across-family discussion.

The initial chapters are about the employment of mites against some specific groups of agricultural and medical importance, and another chapter is on the use of mites against stored-product pests. The direct and indirect effects of the host plant on acarine biocontrol agents are then discussed, with the implication that the plant on which both pests and predaceous mites live is not a passive presence. This is followed by a chapter on the direct and indirect effects of pesticides on the various predators, and another on the rearing and shipment of these beneficial arthropods. The first of the next two chapters tries to supply guidelines about the demonstration of the efficacy of acarine biocontrol agents, whereas the second discusses the attributes of the more efficient amongst these mites. We close with a list of recommendations, again supplementing those presented by the contributors to Hoy *et al.* (1983).

32
Acari as natural enemies of nematodes

The Acari feeding on nematodes were reviewed by Small (1988) within the context of other invertebrate predators. Mites belonging to various suborders feed on pest nematodes to the extent of affecting their populations. These include the Astigmata (Muraoka and Ishibashi, 1976; Sturhan and Hampel, 1977), the Cryptostigmata (Rockett, 1980) and the Mesostigmata (Imbriani and Mankau, 1983; Karg, 1983; van de Bund, 1972). Several of these authors also reported on small-scale, pot experiments which indicated that mites brought about considerable reductions in nematode numbers. Sharma (1971), for example, showed that in pot experiments the ascid *Lasioseius penicilliger* Berlese reduced the populations of a plant nematode by 44%, the laelapid *Hypoaspis aculeifer* (Canestrini) by 68% and the rhodacarid *Rhodacarus roseus* Oudemans by 85%. Mankau (1980), reviewing control of nematode pests by natural enemies, believed that mites, which have greater searching ability than other predators, should be further investigated, despite some drawbacks, such as non-specificity, being limited to the upper soil layer, which is only a portion of the vertical range of some pest nematodes, restriction by size to soils with pore spaces which allow for their passage, and initiation of attack only by fortuitously coming upon the nematodes. Walter *et al.* (1988), however, postulated that some of the smaller mesostigmatid mites, with narrow and flattened bodies and suitable mouthparts, probably have access to most of the soil pore space available to nematodes. Furthermore, counts of mite populations conducted in various north European soils by Karg (1983) indicated mean averages of 37 500–370 000/m^2 nematophagous Acari down to a depth of 15 cm. This author believed that under certain conditions predatory mites could protect field crops from nematode damage.

It is evident that the time has come to conduct appropriate field experiments in order to evaluate the significance of various mites, indigenous as well as introduced, in the control of pestiferous nematodes. The recent reports that certain prostigmatids are obligatory nematophages (Walter, 1988), and that consumption by predatory mites corresponded to 20–60% of estimated nematode production in Swedish soils (Lagerlof and Andren, 1988), add weight to this suggestion.

33
Mites which affect grasshopper and locust populations

Members of several mite families attack grasshoppers and locusts (Orthoptera: Acrididae), but only those of the Podapolipidae and Trombidiidae appear to have any economic importance (Eickwort, 1983; Welbourn, 1983). Mites of the former family are minute and often overlooked, whereas those of the latter group are large, red and easily seen. Some species of the podapolipid genus *Locustacarus* live in the large tracheae of various acridids. They are known in the USA, Kenya and New Zealand (Husband, 1974), and there is some indication that their feeding may be slightly detrimental to host health. *Podapolipoides grassi* Berlese is apparently widespread and commonly found on *Locusta migratoria* (L.) and other grasshoppers. They infest the third-instar nymphs and subsequent stages, and usually occur on the thorax or anterior abdominal segments, often inhabiting the bases of the wings. These mites may also attack the genitalia (Gauchat, 1972). The entire life cycle is spent on the locust, and apparently the mites are dispersed during host mating. Heavy mite infestations on laboratory cultures of the Australian plague locust (*Chortoicetes terminifera* Walker) caused a decrease in host vigour. Field populations of *C. terminifera* were also found to be heavily infested with mites, but the effects of the mites on this and other hosts are not known (Gauchat, 1972).

Eutrombidium locustarum (Walsh) is a parasite during its larval stage and a predator as a nymph and adult on many North American species of grasshoppers (Welbourn, 1983). Mite larvae parasitize grasshoppers from their first nymphal instar on (at the rate of about 35 mites/host) with some grasshopper species being preferred (with up to 75% parasitized) over others (Huggans and Blickenstaff, 1966). Mite feeding apparently interferes with normal adult flight by causing grasshopper wings to become battered and even broken (Severin, 1944). Mite nymphs and adults seek out grasshopper eggs in the soil and devour them; a pair of mites consumed about five eggs/day. Each female lays about 4700 eggs, and each generation requires 60 days to complete in the laboratory (Severin, 1944). *Dinothrombium pandorae* (Newell and Tevis), whose larvae parasitize grasshoppers in California and eastern Africa (Newell, 1979), produces over 80 000 eggs/female; nymphs and adults feed on termites (Tevis and Newell, 1962).

Lawrence (1940) briefly reviewed the mite parasites found on locusts and grasshoppers (although without any quantitative data), and cleared up some taxonomic misunderstandings found in the older literature. Very little is known about acarine natural enemies of grasshoppers and locusts at their African breeding sites, and without additional quantitative, as well as qualitative data, no progress can be made in the evaluation of the potential of these mites against the locust plague.

34
Acari as natural enemies of scale insects

Scale insects (Homoptera: Coccoidea) are notorious plant pests, inflicting their damage by feeding, injecting toxins, transmitting diseases and excreting honeydew on which disfiguring sooty-mould fungi grow. These pests injure many fruit, forest and ornamental trees (Miller and Kosztarab, 1979). The most important families are the Diaspididae (armoured scale insects), Coccidae (soft scale), Pseudococcidae (mealybugs) and Margarodidae. The mite predators of the Diaspididae were reviewed by Gerson et al. *Hemisarcoptes* (Hemisarcoptidae) is the best-known acarine natural enemy of armoured scale insects, the others being members of the Camerobiidae, Cheyletidae and Eupalopsellidae; their main attributes (and those of other families of minor significance) were discussed in the appropriate family chapters.

No specific acarine predators of soft scales or mealybugs seem to have been recorded. Ragusa and Swirski (1977) offered eggs and crawlers of these two families to the phytoseiid *Amblyseius swirskii* Athias-Henriot, and reported that these were inadequate diets. The addition of honeydew promoted some mating, but fecundity was still negligible. Nevertheless, honeydew excreted by scale insects does serve as a supplementary diet for many phytoseiids, promoting their survival under harsh field conditions. Several tydeids are often found in association with soft scales, feeding on their honeydew (Brickhill, 1958). One tydeid species feeds on this diet to such an extent that it is a sanitizing agent in citrus groves, eliminating the honeydew substrate on which sooty-mould fungi develop (Mendel and Gerson, 1982), thereby reducing the pests' damage. Other tydeids may well have a similar function.

Tandon and Lal (1976) reported that the erythraeids *Leptus* and *Bochartia* fed on *Drosicha mangiferae* Green (Margarodidae), a mango pest in India. Sucking by these mites (1–60/host) killed the pests of which 15–20% were attacked. The economic impact of these natural enemies was not evaluated. *Pyemotes* spp. was reported by Mallea et al. (1983) to parasitize the soft scale *Saissetia oleae* (Olivier), a citrus pest, in Argentina, but no further information is available.

These reports suggest that additional natural enemies of scale insects remain to be discovered, especially in the warmer parts of the world.

35
Acari as natural enemies of aquatic Diptera of medical importance

The larvae of several dipterous families of medical or veterinary importance develop in water and they or the emerging adults are attacked there by various mites. The flies involved include mosquitoes (Culicidae), biting midges (Ceratopogonidae), buffalo gnats or black flies (Simuliidae) and horse flies (Tabanidae), whereas most of the Acari are water mites (Hydrachnida). The larvae of these mites are usually parasitic on adult insects whereas the nymphs and adults are predators of various aquatic arthropods (including larvae of the pestiferous Diptera). The literature (mostly in regard to mosquitoes) was critically reviewed by Mullen (1975) and more recently by Smith (1983, 1988); water mite diets were summed up by Böttger (1970). Older records (e.g. Abdel-Malek, 1948), while emphasizing the beneficial effect of the 'Hydracarina', often neglected to identify these mites any further. (More recent records may also require vigorous re-determinations according to Smith and Oliver, 1986).

Mites affect these flies by direct predation, and/or by curtailing survival of pre-ovipositing females and/or by reducing fecundity. The differences in size between parasite and host are of paramount importance in the latter cases (especially with mites which engorge rapidly). Small mites hardly affect large hosts, but parasites large relative to their hosts may sometimes be very detrimental (Lanciani, 1983), even totally reducing the host's fecundity (Davids and Schoots, 1975). The nymphs and adults of *Piona nodata* (Muller) are voracious predators of mosquito larvae (Smith, 1983). The larvae of *Thyas barbigera* Viets parasitize adult mosquitoes whereas the nymphs and adults feed on the immatures. Davids and Schoots (1975) and Lanciani (1979a, 1982) demonstrated the influence of water mite parasitism on non-pest insect populations, while Lanciani (1986) and Smith and McIver (1984) showed similar adverse effects on ceratopogonid and mosquito numbers. Davies (1959) suspected that parasitism by a *Sperchon* sp. may help to control the populations of a minor black fly pest. Gledhill *et al.* (1982) suggested that the morphology of black fly pupae could be a major factor in host selection by larvae of a *Sperchon* sp.

There is a clear need for further quantitative studies on these relationships, including observations on predation by the postlarval stages. Lanciani (1983) suggested that field experiments aimed at using water mites as biological control agents of medically important flies would be the most direct way of evaluating these natural enemies. More specifically, Smith and McIver (1984) advocated the introduction of *Arrenurus danbyensis* Mullen against exotic mosquitoes. Such efforts would probably have a higher chance of success under tropical or subtropical conditions, where hosts are available all the year around.

36
Acarine biocontrol agents in stored products

The utilization of mites for the biological control of stored-product pests was reviewed by Bruce (1983), who noted that most of the important agents belonged to the Ascidae, Cheyletidae and Pyemotidae; data on these families were included in the family chapters.

Use of biological control in stored products has often met the objection that, even if the pests were to be totally controlled by natural enemies, contamination would still be caused by the bodies of the latter. This aspect was discussed by Haines (1981), who pointed out that such reluctance occurs in storage and marketing systems where consumer tolerance to 'impurities' is extremely low. In tropical marketing systems, however, such fastidiousness is rare, and the low level of quality contamination caused by a successful natural enemy would be an enormous improvement on the serious losses caused by the pest. It is difficult to decide whether such considerations have played a role in the total lack of introductions of predaceous mites into the stored-product habitat; whatever the reason, this remains an unexplored, promising research area. One potential source for suitable acarine biocontrol agents will be discussed below. Other stored products wherein biological control could be an option are the health foods, whose industries claim that their products are totally 'natural' (grown and preserved without any chemicals, including pesticides).

Lists of stored-product mites, published in many parts of the world (Hughes, 1976; Smiley, 1984) show a surprising similarity in the pest species, which suggests that the microclimatic conditions in storage are rather similar. On the other hand, quite a few of the predators are different. The list published by Tseng (1979) for the Taiwanese stored mites may be compared for this purpose with that presented by Hughes (1976), mostly from Europe (Table 2). Tseng noted 27 pest species, of which 24 (90%) were also mentioned by Hughes. But of the 21 predators listed from Taiwan, Hughes noted only nine (43%). In other words, numerous acarine natural enemies, which are already pre-adapted to the stored-product environment, are available for additional introductions. These, and other acarine predators obtained from the pests' field habitats (Griffiths, 1960; Sinha and Wallace, 1966), should be tried against these pests.

Table 2. Predaceous mites recorded from stored products in Taiwan (Tseng, 1979) compared to their listing by Hughes (1976) (+++ = present; --- = absent)

			Tseng	Hughes
PROSTIGMATA	Bdellidae	*Bdella lignicola* Canestrini		+++
	Cheyletidae	*Caudacheles lieni* Tseng		---
		Cheletomimus binus Tseng		---
		C. bisetosa Tseng		---
		Cheletomorpha lepidopterorum (Shaw)		+++
		Cheletonella pilosa Tseng		---
		Cheyletus eruditus (Schrank)		+++
		C. fortis Oudemans		---
		C. malaccensis Oudemans		+++
		C. trouessarti Oudemans		+++
		Eucheyletia reticulata Cunliffe		---
		Grallacheles bakeri De Leon		---
		Hemicheyletia arecana Tseng		---
	Cunaxidae	*Cunaxa taurus* (Kramer)		+++
		C. womersleyi (Baker and Hoffmann)		---
MESOSTIGMATA	Ascidae	*Lasioseius allii* Chant		---
		L. martini Tseng		---
		L. sugarawai Ehara		---
		Melichares agilis Hering		+++
		M. mali (Oudemans)		+++
		Proctolaelaps pygmaeus (Muller)		+++

37
Influence of host plants on the efficacy of acarine biocontrol agents

Host-plant characteristics are known to influence the performance of natural enemies attacking the pests of these plants (Price *et al.*, 1980), and the same has been shown for predaceous mites. Such effects can be direct (influencing the natural enemy) or indirect (through the prey).

Leaf surface texture and vestiture affect the performance of acarine predators. Downing and Moilliet (1967) compared phytoseiid efficacy in controlling the European red mite on three apple varieties. The predators were more numerous and control was better on the hairy leaves of the varieties McIntosh and Spartan, whose veins are also more pronounced, than on the relatively smooth leaves of Delicious. Another varietal characteristic believed to have caused this differential behaviour was the presence of rough fruit spurs, which provided phytoseiids with more overwintering shelter and better protection from their own enemies. Overmeer and van Zon (1984) studied the preference of the phytoseiid *Amblyseius potentillae* (Garman) for certain plant substrates, and concluded that this predator too will more often be found on pubescent leaf surfaces. The anystid *Anystis baccarum* (L.), on the other hand, provided far better control of spider mites on glabrous berry leaves than on hairy soybean leaves (Lange *et al.*, 1974a). Leaf domatia, minute cavities frequently located at the junctions of primary and secondary veins on the underside of woody dicot leaves, often harbour various predaceous mites (Pemberton and Turner, 1989). The composition of this variable acarine predator guild may well depend on the plant species. Various arrays of alternative prey and/or competitors will be present on different host plants, with variable effects on acarine predators.

It is commonly known that certain predaceous mites prefer one plant habitat over another, some being found mostly on trees and shrubs while others occur exclusively on annual or recumbent plants. Furthermore, different varieties of the same commercial plant will support quite different numbers of predators (Woolhouse and Harmsen, 1984). Such observations, which seldom find their way into the literature, are of obvious applied significance.

Plants bear various structures and produce secretions which impede the movement of small predators (Price et al., 1980; van Haren et al., 1987). Such structures increase the searching time of acarine biocontrol agents, thereby hindering their activities. On the other hand, many plants provide floral as well as extrafloral nectars, pollen and even nutrients from other plant feeders (i.e. homopteran honeydew), natural foods known to be of importance for many general predators (McMurtry and Rodriguez, 1987). The quality (and quantity) of such natural foods differs from plant to plant with obvious effects on the predators; Abo Elghar et al., (1969) demonstrated that a stigmaeid ingesting palm date pollen will produce more than twice as many progeny as one feeding on castor bean pollen. Plants likewise support additional phytophages and saprophages and are susceptible to different fungal diseases. Both groups of organisms (and their secretions) may serve as alternative food and/or as competitors, thus affecting the populations and behaviour of acarine predators.

Another plant influence which may be direct or indirect (see below) is the supply of vitamin A to *A. potentillae* (in the form of β-carotene) via plant pollen (Overmeer and van Zon, 1983). Albino strains of this mite require the vitamin as a receptor for the appropriate photoperiod which will induce their winter diapause.

There are many indirect, via-the-prey, effects of plants on mite predators. Various plant species and their varieties cause different population levels of spider mites (and other pests) to develop (Gerson and Aronowitz, 1980; van de Vrie et al., 1972), and such different prey levels usually engender varying predator numbers. Diverse plants also cause spider mites to produce variable amounts of webbing (Gerson and Aronowitz, 1981), webbing which usually hinders generalist predators (like *Anystis*) while encouraging specialists (i.e. *Phytoseiulus persimilis* Athias-Henriot) (Gerson, 1983).

Agricultural practices, intended to improve plant quality, whether the addition of mineral fertilizers or the development of resistant plant varieties, may likewise affect prey and predator numbers. (The effect of plant protection chemicals is separately discussed.) Hamai and Huffaker (1978) grew strawberries at low and high nitrogen levels and consequently obtained low and high levels of two-spotted spider mite, *Tetranychus urticae* Koch. The phytoseiid *Typhlodromus occidentalis* Nesbitt controlled the pest on both groups of plants, but economic control was less reliable on plants given the high nitrogen treatment. Santos (1984) reported that apple leaf condition, manipulated by different nutrient levels and humidity conditions, affected populations of both the European red mite, *Panonychus ulmi* (Koch) and its predator, the stigmaeid *Zetzellia mali* (Ewing), but the overall predator–prey cycle was not really affected.

Better plant nutrients may also strengthen the prey pests, thereby enabling them to defend themselves better against some acarine attacks, as well as develop faster, thus lessening their exposure to predators.

Plant resistance has not always been considered to be compatible with biological control (for reviews see Bergman and Tingey, 1979; Boethel and Eikenbary, 1986), although such compatibility could well depend on the specific resistance mechanism, whether due to antibiosis, non-preference or tolerance. Fewer pests per plant, toxic allelochemical substances, or even nectarless varieties would all reduce predator populations. Acarine biocontrol agents can be adversely affected through inhibitory plant components ingested along with prey body fluids. Females of *P. persimilis*, offered *T. urticae* reared on nightshade (*Solanum douglasii* Dunal) and on beans, were significantly lighter (and, by implication, less fecund) when fed on prey reared on the former host plant (Moraes and McMurtry, 1987). On the other hand, the prolonged development of pests on resistant plants would lead to greater exposure to predators, and large weakened prey could more easily be overwhelmed by predaceous mites. Specific plant resistance factors have been shown to affect the performance of an efficient acarine predator. A tomato variety with sticky trichomes trapped and killed many *P. persimilis*, and also slowed the spread of any surviving predators (van Haren et al., 1987). This finding has obvious implications for the release strategy of *P. persimilis* on tomatoes and serves as a reminder that host plants are not passive entities in biocontrol projects. Tolerant plant varieties, which can carry numerous pests without suffering discernible economic damage, are probably most suitable for integration in biological control projects.

The importance of carotene of plant origin in helping an albino strain of a phytoseiid to enter diapause has already been noted. The carotene may also be obtained from the spider mite prey, which, in turn, had obtained it from the host plant. The reaction of *A. potentillae* to spider mite pheromones differs according to the diet on which it had been reared. When offered only carotene-free broad bean (*Vicia faba* L.) pollen, the predator was affected by a pheromone from *T. urticae* (which is an inferior prey). However, when fed on a carotene-including diet, *A. potentillae* responded to a pheromone from the superior prey, *P. ulmi* (Dicke et al., 1986).

Some plants have been postulated to 'recruit' predaceous mites as bodyguards, either before or following attack by spider mites (Dicke and Sabelis, 1988). This attractive theory (supported by the aforementioned leaf domatia hypothesis of Pemberton and Turner, 1989) will probably have to be vigorously tested with non-acarine pests and in many field situations, especially on wild plants, before it is widely incorporated in plant breeding programmes.

Acarine predators can be affected by pesticides via the food chain while feeding on plant pests (Daneshvar and Rodriguez, 1975; Lindquist and Wolgamott, 1980). Pesticide detoxification enzymes may also be obtained in the same manner; Strickler and Croft (1985) reported that the level of mixed-function oxidases of a spider mite and its phytoseiid predator were lower in mites (pests and natural enemies) reared on cucumber than when they were cultured on beans. Asano and Kamei (1982) showed that the effect of the acaricide cycloprate (Zardex®) on spider mite eggs differed according to the host plant, eggs on soybean leaves being much less susceptible than those on peach, orange or apple leaves. This effect was attributed to the higher ability of soybean leaves to affect translaminar transport of the chemical. Eggs which contain higher pesticide concentration can be expected to have a greater effect on their predators.

To conclude, we fully concur with Price *et al.* (1980) that 'we cannot understand predator–prey relationships without understanding the role of plants', and urge that more studies be conducted about these interrelationships, especially in regard to predaceous mites. It should, however, always be borne in mind that interactions seen on commercial plants might well differ from those which had evolved on the pests' and the natural enemies' original host plants.

38
The effect of pesticides on acarine biocontrol agents

The contribution of acarine biocontrol agents to pest control was often inferred from outbreaks of phytophagous pests following the inadvertent destruction of their acarine natural enemies by plant protection chemicals (Garman, 1948; Gilliatt, 1935; Lord, 1949). These effects were recognized and analysed in an early review (Ripper, 1956), which included five predaceous mites (belonging to the families Anystidae, Bdellidae, Hemisarcoptidae, Phytoseiidae and Stigmaeidae) in its list of natural enemies whose beneficial activities were curtailed by pesticides. Much of the current interest in the Phytoseiidae stems from the realization that the emergence of spider mites as major plant pests after World War II was caused mostly by the pesticidal reduction of their natural enemies, of which phytoseiids are a major component (Clancy and Pollard, 1952; Lord, 1949).

The search for plant-protection chemicals whose toxicity will be more acute to the pest than to its acarine natural enemies has engendered considerable research. Gilliatt (1935), Lord (1947, 1949) and Lord and MacPhee (1953) were among the first to assay the field effects of spray programmes on these mites. Their finding that many common pesticides, including sulphur and winter oils, were detrimental to phytoseiids and to *Hemisarcoptes malus* (Shimer) helped explain outbreaks of the European red mite and the oystershell scale (*Lepidosaphes ulmi* (L.)) on apples in Nova Scotia. Such work was later extended to other acarine predators (Herne and Putman, 1966; MacPhee and Sanford, 1954; Parent, 1961). Ristich (1956) apparently conducted the first large-scale laboratory screening experiment on the effects of many (25) commercial pesticides on a single phytoseiid species (see Bartlett, 1964 for an early review). World-wide recognition of the importance of studying the effects of pesticides (including fungicides and herbicides) on biocontrol agents has produced continuing international programmes for appropriate testing (Hassan *et al.*, 1988). Several phytoseiids are among beneficial organisms whose survival rates after exposure to plant protection chemicals are monitored by standard pesticide test methods.

The following brief summary of the effects of insecticides and acaricides on acarine biocontrol agents (presented alphabetically but excluding mites which feed on weeds; note that most of these Acari belong to families that contain pests against which the acaricides were developed) is intended to show the extent of on-going research and makes no claims at comprehensiveness. Insecticide groups rather than individual chemicals are usually noted, except for items of special interest. Data were obtained from four major sources:

1. Field experiments in chemical pest control, whose results indicate side effects on predaceous mites (Graham, 1970; Lord, 1949; Herne and Putman, 1966).
2. Field experiments intended to demonstrate the influence of pesticides on beneficial mites, or to explore the latters' efficacy (Herne and Putman, 1966; Mendel and Gerson, 1982; Parent, 1961; Swift, 1970).
3. Laboratory investigations on the effects of pesticides on predaceous mites (Bartlett, 1964; Hassan *et al.*, 1988; Royalty and Perring, 1987).
4. Studies on the chemical control of mites which disrupt mass rearings of insect pests in laboratory cultures (Sellers and Robinson, 1950; Strong *et al.*, 1959; Stein, 1960).

Anystidae   Carbamates, chlorinated hydrocarbons (CHs) and organophosphores (OPs) were very toxic, although the OP diazinon seemed to be relatively harmless (Herne and Putman, 1966; MacPhee and Sanford, 1954).

Ascidae   Malathion (an OP) was very toxic to *Blattisocius tarsalis* (Berlese), with a consequent outbreak of the stored-product pest *Ephestia cautella* (Graham, 1970).

Bdellidae   The extreme sensitivity of these mites to HCs was used by Wallace (1954) to demonstrate their efficacy in the field.

Cheyletidae   The acaricides chlorobenzilate, dicofol and chloropropylate were lethal for these mites, while Tedion and amitraz allowed some some survival. Synthetic pyrethroids (SPs) were very harmful and the effects of OPs were mixed (Avidov *et al.*, 1968; Stein, 1960; Strong *et al.*, 1959; Žďárková and Horak, 1987).

Erythraeidae   CHs, OPs, SPs, and sulphur were all quite lethal to *Balaustium* (Hagley and Simpson, 1983; Herne and Putnam, 1966).

Hemisarcoptidae   *H. malus* was susceptible to sulphur and winter oils and an OP, and there were mixed results in regard to CHs and arsenates (Lord and MacPhee, 1953; MacPhee and Sanford, 1954). When these mites became pests of an armoured scale insect culture, Sellers and Robinson (1950) eliminated them with Neotran® (a CH).

Macrochelidae Most CHs and OPs were very lethal to these mites (Axtell, 1966), but the insect growth regulator cyromazine, applied against house flies, had no adverse effect on Macrochelidae (Axtell and Edwards, 1983).

Phytoseiidae No group of predaceous mites has been studied in greater detail concerning their reactions to pesticides than have members of the Phytoseiidae. Extensive lists of the toxicities of many pesticides (including botanicals, oils and fungicides) to individual species (i.e. Bartlett, 1964; Croft and Brown, 1975; Hassan et al., 1988 and former papers) indicate that CHs, acaricides and some OPs (such as diazinon) are somewhat less toxic than other OPs, although much variation in these results was noted. The newer SPs are very toxic to phytoseiids (Gerson and Cohen, 1989, Penman et al., 1981). The discovery and utilization of pesticide resistance in some phytoseiids has led to numerous pest control programmes with resistant predators at their core (Hoy, 1985).

Pterygosomatidae These mites are much more tolerant towards OPs and carbamates than other external acarine parasites (Lehmann, 1970).

Stigmaeidae Members of this family appear to be tolerant to the CH DDT, to most OPs (except parathion and endosulfan), but were affected by acaricides, carbamates and SPs (Collyer, 1964a; Motoyama et al., 1970; Muma and Selhime, 1971; Nelson et al., 1973; Parent, 1961; Weires and Smith, 1978).

Tydeidae The acaricides cyhexatin and dicofol were very toxic (Mendel and Gerson, 1982; Knop and Hoy, 1983), while avermectin B_1 appeared to be less lethal to these mites. The latter compound may, therefore, be used selectively against an eriophyid pest of tomatoes, without reducing the numbers of a predaceous tydeid (Royalty and Perring, 1987).

Fungicides have long been known to be more lethal to mites than to insect pests (Dean et al., 1966). Sulphur treatments applied against apple scab (*Venturia inequalis* Wint.) in Nova Scotia destroyed the natural enemies (*H. malus* and *Aphytis*, an aphelinid parasite) of the oystershell scale and brought about an outbreak of the pest (Lord, 1947). Sulphur is very detrimental to predaceous mites belonging to various families

(Bartlett, 1964; MacPhee and Sanford, 1954; Muma and Selhime, 1971). Zineb is almost harmless to Cheyletidae and Stigmaeidae (Avidov et al., 1968; Bartlett, 1964; Muma and Selhime, 1971). Hassan et al. (1988) recently showed that of a group of eight modern fungicides, one (metiram) was lethal to phytoseiids, another (propiconazole) moderately harmful, and the rest had no effect. These results (as well as data presented by van Zon and Wysoki, 1978) reflect the diversity of fungicide effects on predaceous mites. The variable consequences of fungicide use on phytoseiid populations in the field will be noted below.

Herbicides, usually of little or no effect on beneficial arthropods, may sometimes be detrimental to predaceous mites. Of four herbicides tested by Hassan et al. (1988), one (bromoxynil) was harmful to three out of four phytoseiids assayed. Pfeiffer (1986) showed that paraquat and glyphosate reduced phytoseiid numbers in the orchard, with concomitant increases in pest spider mite populations. This may have been caused by direct toxicity (Rock and Yeargen, 1973) and/or by reducing the amount of weed cover in which the phytoseiids overwinter (and which may supply pollen as well as alternative prey).

Much variability has been reported in regard to the toxicity of pesticides to predaceous mites. This variability probably reflects differences in formulation and application techniques, i.e. dusting vs. spraying, low vs. high volume spraying (White and Laing, 1977a), in target-mite stage (Stein, 1960; van Zon and Wysoki, 1978), in specific mode of action (even of pesticides belonging to the same group; the rather mild influence of the OP diazinon, as noted above, is one example, and the different effects of SPs on mites within the suborder Prostigmata (Gerson and Cohen, 1989) is another), in their persistence (Bartlett, 1964), and in different laboratory test methods (especially in the past). Contributing factors are mistakes in identifying experimental animals, the increasing tolerance (i.e. as based solely on field observations) and resistance (i.e. as based on standard laboratory experiments) of various predatory mite strains to agricultural chemicals, and, in regard to field data, meteorological phenomena and effects of the host plant (Chapter 37). The habits of the predators can also expect to affect their sensitivity: more-active mites (such as the Phytoseiidae) will be more likely than sedate ones (i.e. Cheyletidae) to come in contact with the toxicant, and those that supplement their diets by sucking on the host plant (i.e. certain phytoseiids) are more sensitive to systemic pesticides than those with strictly carnivorous habits (Congdon and Tanigoshi, 1983). Phytoseiids were strongly repelled by sublethal doses of various pesticide residues (Hislop et al., 1981; Penman et al., 1981), adding yet more variability, because not all pesticides elicit this behaviour from predaceous mites.

EFFECT OF PESTICIDES

Another factor contributing to the variable effect of pesticides is the food-chain factor: the OP acephate was very toxic to *P. persimilis* preying on *T. urticae* which had become insecticide-contaminated after the pesticide was applied as a soil drench (Lindquist and Wolgamott, 1980). On the other hand, Congdon and Tanigoshi (1983), who reviewed the subject, found that a phytoseiid which had fed on the pesticide-contaminated citrus thrips (*Scirtothrips citri* (Moulton)) suffered no mortality. The secondary chemicals of some plants may affect the susceptibility of pests and their natural acarine enemies to certain pesticides. Strickler and Croft (1985) reported that the LC_{50} of rotenone for *Amblyseius fallacis* Garman was considerably different when fed *T. urticae* reared on two hosts: cucumbers and beans.

Population increases of mite-controlled pests following a pesticide application are usually attributed to the elimination of acarine natural enemies. This 'elimination' may actually consist of a number of processes, and their understanding could well contribute to better-integrated pest control with mites. (For the sake of simplicity the effects of pesticides on herbivorous competitors are not considered in the following discussion.) In the simplest case, the pesticide kills the natural enemy while sparing the pest (whether due to insensitivity or resistance). In other cases, the slight or pronounced mortality of natural enemies causes pest numbers to increase above the economic level. Lord (1947) found that 'mild sulphur' sprays (like flotation sulphur) killed the natural enemies of the oystershell scale while hardly affecting the pest, resulting in the latter's outbreaks. Stronger sulphur (lime sulphur), however, while still being lethal to the natural enemies, also brought about some scale control. Nevertheless, the slight advantage thus accruing to the pest resulted in its outbreaks, albeit less rapidly. Another possibility is that the toxicant kills all pests as well as all predators, but the former, whose only requirement for food is the plant, re-colonize this substrate more quickly than their enemies (which need a certain minimal number of prey pests in order to survive on the plant), still bringing about pesticide-induced economic injury. The emergence of pesticide tolerance and resistance (two phenomena often confused in the literature) in phytoseiids and other predaceous mites (as listed above) tends to offset some of these effects. Certain pesticides (especially SPs) are not only very toxic to many predators but, independently, also cause resurgences of spider mites, engendering even larger pest populations (Gerson and Cohen, 1989).

The rise of pesticide-resistant phytoseiids is bringing about other subtle, little-studied phenomena, such as the displacement of resident, possibly better-adapted predators by resistant ones. OP sprays caused the disappearance of *Typhlodromus caudiglans* Schuster and *T. pyri*

Scheuten from treated apple trees; a resistant strain of *T. occidentalis* was then introduced. Later, when the sprays were stopped, the resident phytoseiids re-emerged, relegating *T. occidentalis* to a minor status (Downing and Moilliet, 1972). The fungicides benomyl and fentin hydroxide caused high but differential mortality to three species of phytoseiids inhabiting pecan foliage in Florida (Ball, 1982). The most abundant species was almost eliminated, while another became dominant on the treated trees.

Current research in this field (apart from that noted above) includes the continuous genetic improvement of phytoseiids (Hoy, 1985), selection of strains of two successful members of this family for resistance to novel insecticides in New Zealand (Markwick, 1986), and the continuing search for selective toxicities of new pest control chemicals. Grafton-Cardwell and Hoy (1983) reported that avermectin B_1 was sufficiently less toxic to a phytoseiid, as compared to two spider mite species, that some pest management was possible while using this new pesticide. Comparable results were reported by Osborne and Petitt (1985) with Safer's Insecticidal Soap. Pronounced differences in favour of the predator were obtained by Mansour *et al.* (1987) in regard to neem, an insecticide extracted from the seed kernels of *Azadirachta indica* A. Juss. On the other hand, some of the recently introduced insect growth regulators (IGRs) seem to bring about spider mite outbreaks, possibly through their adverse effects on phytoseiid predators (F. Mansour, personal communication). Bellows and Morse (1988 and former publications) are monitoring the effects of pesticide residues on various natural enemies of citrus pests in California, including one phytoseiid mite.

The effects of behavioural resistance (of pests and of their mite predators) and of synergists (and other pesticide additives) are two additional areas which could be worthy of further research effort. Lockwood *et al.* (1984) discussed the difficulties of demonstrating behavioural resistance to pesticides; nevertheless, such studies could be of much value in the present context. Synergists can be used as research tools to define the potential toxicity of various pesticides and may aid in determining various mechanisms of resistance (Raffa and Priester, 1985). Use of appropriate synergists could also render pests more, and predaceous mites less, susceptible to some pesticides.

39
Rearing and shipping

In order to obtain sufficient numbers for field releases, acarine biocontrol agents may be either collected in the field or mass-reared in the laboratory, or both. When mites are then shipped to release sites, special precautions must be taken to maintain optimal humidities in the containers as well as sufficient aeration. Collected mites can be kept for different periods under low temperature conditions (e.g. Žďárková and Pulpan, 1973). A prerequisite for the development of mass rearing and shipment methods is information on the life histories and specific feeding habits of these mites (Barker, 1968; McMurtry and Rodriguez, 1987; Rodriguez et al., 1962 and many others).

Currie (1934) provided an early example of the field collection of a predaceous mite, the bdellid *Bdellodes lapidaria* Kramer. Observations had shown that mites hid under bark or fallen branches. Boards were then placed in the field as mite 'traps', and two collectors, using aspirators, subsequently obtained 1000 mites/person/hour. This method was subsequently widely used, resulting in a collection of 14 750 *Bdellodes*. Working in another era, Wallace (1972) developed a portable, motor-operated sucking machine to collect large quantities of these mites in the field. Lange et al. (1974b) employed three different methods in order to obtain large numbers of *Anystis baccarum* for spider mite control. They sieved out egg clusters from the pine and oak litter wherein they had been deposited, collected moving mites with a sweepnet provided with a collecting bottle at its bottom, and trapped adults which were ascending or descending trees with belts open only at the entrance. Predators obtained by either of the two latter methods were then immediately used for pest control.

Rangel (1901) was probably the first to rear a beneficial mite in the laboratory, namely *Pyemotes* sp., assayed against the cotton boll weevil (*Anthomonus grandis* Boheman) in Mexico at the turn of the century. Weevil-infested bolls were brought into the laboratory, air-dried (to protect against fungi) and exposed to the mites. Rangel (1901) was also the first to suggest the use of alternative hosts (in this case, developmental stages of a house-dwelling wasp) for the mass-production of an acarine biological agent. Methods of mass-rearing *Pyemotes* on insect hosts were later described by Bruce (1983) and by Weiser (1963). Washed moth, beetle or ant larvae or pupae were exposed to host-seeking female mites, which rapidly colonized the insects and raised a generation in one week at 25°C. Bruce (1989) more recently devised an artificial medium for *Pyemotes*.

Although the biology of the cheyletid *Cheyletus eruditus* (Schrank) is fairly well known (Solomon, 1969), it may be easier to obtain this mite in large numbers by sieving it out of infested substrates (Pulpan and Verner, 1965). Or, if mass-reared, it can still be separated from wheat by this or by non-lethal flotation methods.

Hemisarcoptes coccophagus Meyer was reared in the laboratory on armoured scale insects maintained on potato tubers at 80% relative humidity (Gerson, 1967). Cultures were begun by placing incised hypopus-bearing elytra of the coccinellid beetle, *Chilocorus bipustulatus* (L.), the specific vector, on the tubers. Emerging tritonymhs then immediately attacked the scales and continued their life cycles.

Macrochelids were routinely mass-reared on a medium formerly used for house flies, to which nematodes were added (Willis and Axtell, 1968). Another predator of the house fly larvae, the uropodid *Fuscuropoda vegetans* (De Geer) was cultured on the same diet.

Some phytoseiids are currently being mass-reared on a commercial basis, the exact methods (including plant substrate and variety as well as predator seeding and harvesting) being at times in the realm of 'trade secrets'. Overmeer (1985) reviewed various rearing methods of mites of this family, including on artificial substrata and detached leaves. Hoy *et al*. (1982) described a method for the large-scale rearing of phytoseiids in the field during summer.

At times it becomes more convenient to rear predaceous mites on alternative prey. Two species of *Amblyseius* (Phytoseiidae), mass-produced and released as natural enemies of *Thrips tabaci* Lindeman, were reared on an acarid mite cultured on wheat bran (Ramakers, 1983). Rasmy *et al*. (1987) used a similar diet to rear the stigmaeid *Agistemus exsertus* Gonzalez, for use in spider mite control. The development of artificial diets could be a major step in simplifying mass-production of predaceous mites. This was recently achieved by Ochieng *et al*. (1987), who reared *Amblyseius teke* Pritchard and Baker (employed for control of the cassava green mite, *Mononychellus tanajioa* (Bondar)) for 25 generations on an artificial diet. However, predaceous mites reared on unnatural diets in the laboratory may behave in unexpected ways in the field. This could be due to different hunting and feeding behaviour patterns in the two habitats, to a decline in the stability of the predator–prey association during prolonged laboratory culturing, or to contamination by closely related, look-alike species (and to combinations thereof). Regular quality control tests are therefore strongly recommended.

Regrettably, no data are available about the method used by Riley to ship '*Tyroglyphus phylloxerae*' Riley to Europe in 1873, in the first-ever international shipment of an acarine natural enemy (Howard, 1930). Other early efforts are also vague. When *Pyemotes* was introduced into the USA from Mexico in 1902 in order to try it against the cotton boll weevil, the introducer 'brought with him . . . a supply of the parasites', with no further details (Hunter and Hinds, 1904).

Currie (1934) developed a method for sending bdellids from Western Australia to its eastern territories, including Tasmania. A glass tube, open at either end, was plugged at both ends with sphagnum moss wrapped in cellophane. Strips of dry bark and clover leaves were put into the tubes along with the bdellids, as well as prey (collembolans). The sphagnum, moist at one end, dry at the other, served to regulate the humidity inside the containers. Parcels of mite-containing tubes were sent by train and by air and arrived at their destination in good condition. Wallace and Walters (1974), upon shipping Australian-collected bdellids to South Africa, replaced the sphagnum moss with saturated facial tissues, in order to improve aeration and avoid water condensation within the containers.

Hemisarcoptes malus was shipped to Bermuda by Bedford (1949) in the hypopodal stage, under the elytra of *Chilocorus* spp. The same method was recently used while introducing several species of *Hemisarcoptes* into New Zealand (M.G. Hill, unpublished).

Wallace and Holm (1983) surface-sterilized the eggs of *Macrocheles peregrinus* Krantz prior to their shipment by air from South Africa to Australia, where they were tried against flies.

Overmeer (1985) briefly discussed handling and shipment procedures for phytoseiids, which probably include commercial methods. Herren *et al.* (1987) and Pickett *et al.* (1987) are developing systems for the aerial release of phytoseiids intended to control cassava green mite in Africa and corn mites in the USA, respectively. A special feature of Herren *et al.*'s technique is the marking of released predators with fluorescent powders (which persist for about 24 hours), thereby enabling investigators to follow the spread of the mites.

Special care must be exercised when phytophagous mites are to be introduced for weed control (Hill and Stone, 1985). One of the major criteria in choosing a potential weed feeder is the latter's specificity, but this implies that the weed may also have to be introduced for the relevant tests, which in turn necessitates strict quarantine conditions. The problem is especially severe with eriophyiid mites, as they readily disperse on air currents. Rosenthal (1983) described the construction of a special cage used for introducing *Aceria convolvuli* (Nalepa) for the control of field bindweed, *Convolvulus arvensis* L., from Europe into the USA. The cage (devised by D.M. Maddox) prevented mite escape by means of a continuous air flow, away from its door and through a filter fine enough to trap eriophyiids at the rear of the cage. Another option is to transport these mites during hibernation, as Kovalev (1973) suggested in regard to an *Aceria* sp. that occurs, in the thousands, within flower heads of the weed *Acroptilon repens* (L.) DC.

Additional information which may sometimes be required for the optimal rearing, shipping and utilization of predaceous and parasitic mites pertains to the extent of the latter's intraspecific variability. This may be manifested by the better development of certain mite 'races' or 'strains' on some of the pest (including weed) populations than on others, by different rates of sensitivity to environmental conditions (including pesticides) and by partial reproductive incompatibilities. Drummond *et al.* (1985) concluded that the Mexican form of the podapolipid *Chrysomelobia labidomerae* Eickwort represented a different race from the one found in the USA, where it did not occur on the Colorado potato beetle. The Mexican mite race may, therefore, be assayed against the pest. Sobhian and Andres (1978) reported that various European strains of *Aceria chondrillae* (G. Canestrini), introduced into the USA to control the rush skeletonweed, *Chondrilla juncea* L., behaved differently on various strains of the host weed.

Hoy (1982) reviewed intraspecific variability in the Phytoseiidae, listing differences in regard to diapause induction, survival at low temperatures, responses to relative humidities and possibly even to pollen feeding. Information on variability in resistance to pesticides is especially important in biocontrol projects, as are data on partial intraspecific reproductive incompatibilities (Hoy, 1985). Some of this variability probably represents the mite's response to natural or directed selection pressure exerted by new host plants, exposure to agricultural chemicals, mass-rearing conditions and other man-made factors.

It goes almost without saying that correct mite determinations must continuously be at hand from early discovery through life-history studies to mass rearing, field releases and subsequent recovery, and the preservation of voucher specimens (including those of the target-organism) throughout these efforts is strongly recommended. Smiley and Knutson (1983) discussed sources, resources and needs relative to taxonomic research of acarine biocontrol agents.

40
Demonstrating the efficacy of acarine biocontrol agents

The mite families discussed in this book were chosen for inclusion on the basis of claims that some of their members have an adverse effect on pest (or weed) populations. Such claims were at times based on field or laboratory observations alone, unsubstantiated by quantitative data. *Neophyllobius* sp., for instance, was postulated as playing an important role in the natural control of a scale insect in New Zealand (Richards, 1962), and Sturhan and Hampel (1977), on the basis of laboratory feeding experiments and co-occurrence in the field, suggested a role for *Rhizoglyphus echinopus* (Fumouze and Robin) in regulating nematode populations in the soil. Quantitative data on the predators' effect on pest populations in their natural habitats were lacking in both cases (as well as in many others). These data are required in order to establish the status of a given mite predator as a natural enemy capable of regulating pest populations. Such conclusions may be arrived at by experimental methods, by inferable (or indirect) means, or by both.

Luck *et al.* (1988) suggested six general experimental approaches (or techniques) for evaluating the impact of natural enemies on pest populations. These consist of (1) introduction and augmentation; (2) use of cages and barriers; (3) removal of natural enemies; (4) prey enrichment; (5) direct observations and (6) chemical evidence of natural enemy feeding. Luck *et al.* (1988) also provided suggestions as to which techniques would be most suitable relative to the specific objectives of a biocontrol study.

1. **Introduction and augmentation** *Phytoseiulus persimilis* has been introduced into many countries for the control of spider mites infesting greenhouse crops; the increasing use of this predator (van Lenteren and Woets, 1988) attests to its commercial-level efficacy. Experimental evidence for the ability of additional introduced phytoseiids to control pests was reviewed by McMurtry (1982). An example of a non-quantitative report claiming success of an introduced acarine natural enemy was provided by Glendenning (1931), who reported that

Hemisarcoptes malus, introduced into British Columbia from Eastern Canada in order to control oystershell scale 'has effected excellent control'. Experimental proof that the introduction of a natural enemy did not affect pest populations was presented by Thorvilson *et al.* (1987), who reported that field-released *Pyemotes tritici* (Lagreze-Fossat and Montane) failed to control populations of red imported fire ants (*Solenopsis invicta* Buren). Quantitative reports of failures in biological control, which usually go unpublished, are important in laying the foundations for further research. The commercial utilization of pesticide-resistant phytoseiids (Hoy, 1985) are well-documented cases of augmentation.

2. **Cages and barriers** The application of some of these techniques to acarine predators was discussed by Fleschner (1958). Experiments in which numerical changes in citrus spider mites were compared between branches either open or enclosed within organdy sleeves resulted in greater pest increases on enclosed branches. These increases were attributed to microclimatic changes within the enclosure and to restricted pest mite dispersal, rather than to predator action. The side effects of enclosure may thus partially or completely mask the impact of acarine (and other) predators, and the method was not recommended for use with predaceous mites. Confining predators and prey together in cages is another often-used method. Axtell (1963) introduced known quantities of house fly larvae and *Macrocheles muscaedomesticae* (Scopoli) and *Glyptholaspis confusa* (Foa) eggs to fresh manure in cages. By monitoring numbers of emerging flies, Axtell (1963) demonstrated that the predators were responsible for a 83–94% reduction in pest numbers. Imbriani and Mankau (1983) introduced the mesostigmatid *Lasioseius scapulatus* Kennett into nematode cultures and reported that the predator drastically reduced nematode populations.

Caging experiments may show elevated rates of predation, but such results could be unrealistic, due to microclimatic changes, restricted pest and predator movement and because prey finding is probably far easier than under natural conditions (Luck *et al.*, 1988). Barriers can only be used if the mode of natural enemy movement is known, as barriers will not be very effective against predatory mites that are windborne or are carried about by winged hosts.

3. **Removal of natural enemies** These techniques can be subdivided into insecticidal exclusion and biological (or non-chemical) removal methods. For the present purposes the former group can be further divided into intentional and unintentional exclusion. Wallace (1954) demonstrated the efficacy of the bdellid *Bdellodes lapidaria* in controlling

the lucerne flea, *Sminthurus viridis* (L.), a pest of pastures in Australia, by applying the CH pesticide DDT (which has only a limited effect on the pest), and subsequently observing 5- to 19-fold increases in pest numbers. This was considered proof of the bdellid's ability to control the pest. Axtell (1963) destroyed predaceous Macrochelidae with the acaricide Kelthane®, which resulted in significantly reduced house fly control in calf-pen manure as compared to untreated manure. Croft and Brown (1975) discussed the effects of various selective acaricides on stigmaeid and phytoseiid predators and advocated their use as experimental tools for assessing the individual contribution of components of a natural enemy complex. Use of pesticides in the check method may, however, introduce unexpected side effects, as some insecticides are known to elevate spider mite populations, thereby detracting from the method's efficacy (Fleschner, 1958); side effects of pesticides on acarine pests and their predators were discussed above. Unintentional exclusion with subsequent demonstration of the importance of an acarine natural enemy in controlling a pest occurs as a side effect of pesticide usage. Sprays of the OP malathion applied to stored products in Kenya caused an outbreak of the moth *Ephestia cautella*, presumably because the pesticide 'controlled' the pest's most important predator there, *Blattisocius tarsalis*, without seriously harming the pest (Graham, 1970). The importance of the natural biocontrol agent was thus fortuitously discovered.

Biological (i.e. non-chemical) removal of acarine natural enemies can be achieved (or may come about) by various methods, three of which will be noted. Fleschner (1958) compared the efficacy of various methods used to study the effect of natural enemies on spider mites. He stated that hand removal was the most effective and dependable method, as it avoids introducing new factors into the system. But it was also extremely tedious and time-consuming. Another method discussed by Fleschner (1958) was the use of ants, which may inhibit the movement of mite predators. The efficacy of this method depends not only on the ant species present but also on the natural enemies being studied. Haney *et al.* (1987), for instance, showed that densities of a phytoseiid predator were unaffected by the presence of the Argentine ant, *Iridomyrmex humilis* (Mayr), on Californian citrus trees. More suitable antagonists of predaceous mites could, however, be used to demonstrate the efficacy of such natural enemies.

An experiment in which the importance of an acarine natural enemy was unintentionally shown by biological means was reported by Wassenaar (1988). This author vacuumed a cotton carpet several times and then, after seven weeks, cleaned it thoroughly by spray extraction; this was intended to assess the effect of various treatments on house dust mites. Numbers of the latter did not significantly decrease during the course of the vacuuming, whereas those of the predator *Cheyletus* sp. declined. Seven weeks later the populations of the pest had increased enormously, an increase attributed to the destruction of *Cheyletus* by vacuuming.

4. **Prey enrichment** This method which enables the experimenter to control the number of the prey (or hosts), was used by Axtell (1963) in order to demonstrate the effect of macrochelids on filth fly populations. He inoculated fresh cow manure with known numbers of fly eggs and, by combining this technique with insecticidal removal, obtained an estimate of predator efficacy in the field. Ryba *et al.* (1987) tested the ability of various laelapids (Mesostigmata) to affect flea numbers in mammal nests by introducing fleas to all experimental nests and mites only to some. Significantly lower flea numbers were collected from the mite-containing nests.

5. **Direct observations** Luck *et al.* (1988) considered direct observations to be very useful for determining predation rates and identifying the natural enemies. Observations also serve to determine what actually happens when natural enemies meet their prey or host. These methods are especially useful during weed control. Dodd (in Hill and Stone, 1985) stated that *Tetranychus desertorum* (=*opuntiae*) Banks caused a thinning of prickly pear (*Opuntia*) stands in Australia. *Aceria chondrillae* damages rush skeletonweed by forming galls which cause plant stunting and weakening. These galls can be observed; their presence on the weed when attacked by mites of Italian origin, as compared to lack of galling by mites obtained from Greece, persuaded Sobhian and Andres (1978) to introduce mites of the former provenance. Direct observations have the advantages listed above for hand removal; many of the disadvantages can nowadays be surmounted (at least in the laboratory) by using video equipment (Wharton and Arlian, 1972). Such equipment may also provide information on what predators actually do (albeit usually under artificial conditions), especially their patterns of searching behaviour and alternative prey (including cannibalism) selection. Lanciani and Boyt (1977) collected mosquitoes (*Anopheles crucians* Wiedemann), some parasitized by water mites and others not parasitized, and reared them in the laboratory. Parasitized hosts died sooner and deposited significantly fewer eggs than unparasitized mosquitoes, suggesting that in the field, parasitized females have a high probability of dying before obtaining their first blood meal.

Information regarding prey fed on by mites can also be obtained by direct microscopic examination or by chromatography of the gut contents. The predaceous Astigmata and Cryptostigmata eat by ingesting solid food, whereas the Mesostigmata and Prostigmata (and the parasitic Astigmata) feed by sucking out prey (or host) fluids. Visual identification of food remnants in the gut is therefore feasible only for members of the former two groups. A method of gut content analysis of peat bog cryptostigmatids was discussed by Behan-Pelletier and Hill (1983). Some quantitative difficulties with this method are that recognizable sclerotized arthropod remains may accumulate in the predator's gut throughout its life (thus not being attributable to any discrete period of time unless the predator has been observed for given periods or the mite's age is known), and it may be hard to separate between prey eaten alive, moribund, or dead. Pigments of the prey body can be observed in the guts of most acarine predators; Putman and Herne (1964) used paper chromatography in order to demonstrate phytoseiid feeding on spider mites.

6. **Chemical evidence of natural enemy feeding** Electrophoresis and subsequent detection of characteristic esterases by histochemical means were used by Murray and Solomon (1978) to detect predator–prey relationships between phytoseiids and spider mites. A prey mite (the European red mite) was detectable within the gut of a single predator (*Typhlodromus pyri*) at least 31 hours after feeding. Lister *et al*. (1987) calculated the attack rates of the mesostigmatid *Gamasellus racovitzai* (Trouessart) on various Antarctic microarthropods by proportional and quantitative analyses of electrophoretic results. They also devised a method capable of estimating time since feeding and composite meal size for individual predators. Although this predator does not feed on pest species, the methodology developed by Lister *et al*. (1987) should be of much practical value. Other quantitative chemical methods for detecting predation in the field were reviewed by Luck *et al*. (1988) and by Sunderland (1988).

The efficacy of acarine natural enemies in controlling pests may be inferred (or at least suggested) from data obtained in the field, laboratory, or combined observations. At its simplest, the ratios between pest and natural enemy populations in the field, and their co-occurrence (or lack thereof) in the same samples suggest that some interaction (or none) between the two is taking place. Currie (1934) estimated numbers of predaceous bdellid mites and their prey, the lucerne flea, along a line from where mites were very abundant to an area which they had not yet reached. He reported that the pest was far more common in the latter, mite-free areas, and took this as proof of control. Thirty years later Wallace (1967) analysed these interactions and obtained a very significant negative relationship between numbers of bdellids and numbers of lucerne fleas counted 8–9 weeks later. This indicated the importance of predators early in the season. Aeschlimann and Vitou (1986) sampled alfalfa fields infested by aphids in France and found that stations with the trombidiid predator *Allothrombium* had significantly fewer pests than fields without the mite. Inverse numerical relations which developed on apple trees between the phytoseiid *T. pyri* and the European red mite suggested to Collyer (1964b) that the former was affecting populations of the latter. Croft and Nelson (1972) used field data to plot predatory mites against pest mites on apple leaves and produced an index capable of predicting the effect of the natural enemy on the pest at various population densities of either. Santos (1976) reported that the stigmaeid *Zetzellia mali* occurred only on the lower surfaces of apple trees, and primarily on short side branches or spur growth. Its prey, the *P. ulmi* and the eriophyid *Aculus schlechtendali* (Nalepa) inhabited these tree parts as well as the upper leaf surfaces and tips of branches with terminal leaves. Santos (1976) thus concluded that the spatial heterogeneity of the pests was too great for effective control by that predator. However, this conclusion could be misleading under certain circumstances. Nyrop (1988) studied the spatial dynamics of *P. ulmi* and *T. pyri* in the field, and found that the two populations mixed randomly. It was concluded that an acarine predator that had evolved to feed on randomly distributed prey should optimally employ a searching strategy which would result in the same amount of time being spent in all potential prey patches.

Chiang (1970) tried to evaluate the effect of manure applied to corn fields on the feeding of laelapid mites which prey on corn rootworm (*Diabrotica* spp., Coleoptera: Scarabaeidae) in Minnesota. By using elementary mathematical procedures he calculated the number of predaceous mites brought about by the manure, how much *Diabrotica* reduction they caused, how many pests were hitherto destroyed by naturally occurring mites, the number of pests which would have been present without any mites, and hence that the addition of manure caused a 63% reduction in corn rootworm numbers.

Multifactor analyses of life tables were used by Samarasinghe and LeRoux (1966) to demonstrate that *Hemisarcoptes malus* (and an aphelinid wasp) were the key factors regulating the populations of the oystershell scale on apples on Canada. The mite was the sole regulating factor at one locality. Life-table data were employed by White and Laing (1977b) to demonstrate that *Zetzellia mali* was not as effective as some phytoseiids in controlling phytophagous mites.

The effect of water mites on mosquito (*A. crucians* and *Coquillettidia perturbans* (Walker)) populations in the field was investigated by Lanciani (1979a,b) and Smith and McIver (1984), respectively, by inferable methods as well as by direct observations. The latter were used when mosquitoes of different age groups were collected and the number of stylostomes (tube-like structures formed in the host's body when its haemolymph interacts with parasite saliva) counted. Average stylostome number was consistently and significantly higher in young (unfed) as compared to somewhat older *A. crucians*, suggesting that there was a loss of heavily parasitized hosts of the former group (an average of 42.5%, according to a recalculation by Smith and McIver, 1984). Such mortality reduces the number of mosquitoes recruited into the blood-sucking (i.e. pestiferous) population. Lanciani (1979a) also collected engorged *A. crucians* carrying a variety of mite loads and recorded their survival in the laboratory. The instantaneous death rate (Anderson, 1978) of the hosts was then calculated, showing that mosquitoes with higher mite loads survived the least, the adverse effect of parasite load being linear. Lanciani and Boyett (1980) further tried to fit truncated negative binomial distributions (Crofton, 1971) to parasite frequencies within samples of these mosquitoes. Deviations from expected distributions were then used to estimate mite-induced mortalities (especially of the more heavily parasitized mosquitoes). Based on methodology from their own studies, Smith and McIver (1984) calculated that egg production in the first gonotrophic cycle of a parasitized population of *A. crucians* may be reduced by 35%. Additional details of these calculations were presented by Lanciani (1983) and Smith (1988).

The functional responses of acarine predators (changes in number of prey consumed by an individual predator with changes in prey density) and their numerical responses (changes in predator density in response to changes in prey density), were used as indicators of predator efficacy. Laing and Osborn (1974) and Takafuji and Chant (1976) criticized the separate application of each set of responses, arguing that the use of either alone may lead to invalid conclusions. Consequently they advocated the combined use of the functional and numerical responses, and showed how these can be used to explain the efficacy of various phytoseiid predators. Croft and Blyth (1979) investigated both sets of responses of *Amblyseius fallacis* towards the two-spotted spider mite in the laboratory. They concluded that the predator was highly sensitive to changes in prey density and could therefore respond rapidly to spider mite outbreaks in the field.

The effect of the cheyletid *Hemicheyletia bakeri* (Ehara) on populations of the same pest in the laboratory and in the field was studied by Laing (1973). The lack of functional and numerical responses towards different spider mite densities, as well as poor searching ability and limited predator spatial coincidence with the prey, led Laing (1973) to relegate *H. bakeri* to a minor controlling role. Laing and Knop (1983) stated that predators of the families Anystidae, Cheyletidae, Erythraeidae, Stigmaeidae, and Tydeidae are unlikely to control acarine pests because their numerical responses are much lower than those of their prey. Laing and Knop (1983) further stated that this deficiency is not often compensated by a very high functional response. Data presented in the individual family chapters above suggest that this generalization may have been premature. Lister *et al.* (1988) demonstrated that under certain extreme conditions a functional response would have no advantages for the fitness of an acarine predator. Rejection of an acarine predator due to 'limited predator spatial coincidence with prey' could also be misleading, as noted above (Nyrop, 1988).

The many experimental data and inferred conclusions may finally be organized and consolidated to produce predictions of predator efficacy in the field. Dover *et al.* (1979) presented a model which predicts European red mite numbers on apple according to initial pest and predator densities. Logan (1982) and Sabelis (1985) reviewed this and other models with a view to using them to understand phytoseiid behaviour in regard to spider mites, and the potential contributions of such understanding towards furthering pest control in the field.

41
Attributes of efficient acarine biocontrol agents

What makes an effective natural control agent and how can it be recognized without costly and time-consuming tests in the field? This question has often been asked and many efforts have been made to answer it (e.g. Hokkonen, 1985; Huffaker *et al.*, 1976; Turnbull and Chant, 1961; and others). Despite much controversy, the general consensus appears to be that laboratory assays, valuable as they might be, cannot be used with any reasonable certainty to predict the economic-level performance of a given natural enemy in the field. What can and has been done is to study successful cases of biological control and the apparent reasons for their successes, and use any discernible patterns in order to prepare lists of 'desired attributes' of natural enemies.

Rosen and Huffaker (1983) arrived at a list of the 'desired attributes' of effective acarine natural enemies through their analysis of density dependence. Not only were functional and numerical responses considered to be of cardinal importance, but so was the ability of the natural enemy to respond to changes in its own density. They considered the following attributes to indicate the potential value of effective mite predators (subsequently used to denote predators as well as parasites; 'prey' will be employed to mean hosts also).

1. Searching capacity: the ability to locate prey even when they are rare; this is often considered to be the most important attribute of an effective natural enemy.
2. Specificity of natural enemy to prey; a high degree of specificity is considered among the major attributes of a successful natural enemy.
3. Power of increase: the ability of the natural enemy to increase in numbers at least as rapidly as the pest. This attribute, which may be negatively correlated with searching capacity, is especially important in unstable environments.
4. Adaptability of the natural enemy to its environment.

Another effort was made by McMurtry (1982), who compared certain biological characteristics of six phytoseiid species which are known or postulated to be of importance in spider mite control. The attributes compared were: dispersal powers; distribution in relation to prey (roughly comparable to searching ability); reproductive potential (comparable to power of increase); voracity; specificity; and survival ability when prey are scarce (roughly comparable to adaptability).

Summing up his comparison, McMurtry (1982) noted that three phytoseiids, regarded as effective predators, have high dispersal powers, their distributions are closely correlated with those of their prey, they have high reproductive potentials and at least some specificity to spider mites (including an ability to penetrate spider mite colonies protected by dense webbing). On the other hand, they vary in voracity and in their ability to survive prey scarcity. The three other species, considered low on most of these characters, were rated high on their ability to survive low prey availability, and may be important natural enemies under certain circumstances. Results of laboratory studies, even those which vigorously imitate natural conditions and which engender significant data, should thus still be substantiated by

Table 3 Some mites which provide control of agricultural pests

Natural enemy	Pest	Pest order	Habitat	Source
Anystis	Earth mite	Prostigmata	Pastures	Wallace (1981)
	Lucerne flea	Collembola	Pastures	Meyer and Ueckermann (1987)
Bdellodes	Lucerne flea	Collembola	Pastures	Wallace (1954)
Blattisocius	*Ephestia*	Lepidoptera	Stored products	Graham (1970)
Cheyletus	Acarids	Prostigmata	Stored products	Žďárková (1986)
Euseius	Citrus thrips	Thysanoptera	Orchards	Tanigoshi et al. (1984)
Hemisarcoptes	Armoured scale insects	Homoptera	Orchards	Samarasinghe and LeRoux (1966)
Macrocheles	Bushflies	Diptera	Cow dung	Wallace et al. (1979)
Phytoseiulus persimilis	Spider mites	Prostigmata	Greenhouses	McMurtry (1982)
Typhlodromus pyri	Spider mites	Prostigmata	Orchards	McMurtry (1982)
Stigmaeids	*Brevipalpus*	Prostigmata	Tea bushes	Oomen (1982)

field work (McMurtry, 1982). Another important attribute, noted by Schroder (1983), is the ability to tolerate pesticides (whether through insensitivity or by resistance).

As detailed in the family chapters above, some mites were reported to control pest populations and a comparison of certain of their traits could be useful in defining attributes required of an effective acarine biocontrol agent. This exercise will be rather incomplete because (1) only a few successful cases are known and (2) of a lack of basic quantitative information on searching ability (but see Sabelis and Dicke, 1985). Nevertheless, the data (Table 3 and the Family chapters) allow some tentative conclusions to be drawn.

1. Successful acarine biocontrol agents usually possess well-developed searching and dispersal mechanisms; the latter may be autonomous or insect-mediated. Many phytoseiids are fast runners, rapidly 'combing' the immediate environment for food, and/or can be transported by wind; having been blown onto plants, some locate their prey by kairomones (Sabelis and Dicke, 1985). *Anystis agilis* Banks patrols along leaf veins and edges, apparently moving (and searching for prey) in a non-directed way (Sorensen *et al.*, 1976); according to Muma (1975) this species runs about at random, in a figure-of-eight pattern. Bdellids manifest an ambushing strategy and tether their prey (Sorensen *et al.*, 1983); they appear to advance by walking (Currie, 1934). *Blattisocius tarsalis* utilizes stored-product insects for dispersal and then feeds on their eggs (Haines, 1981). *Macrocheles* spp. employ the same strategy, using various filth flies as vectors and food (Axtell and Rutz, 1986). *Cheyletus* spp. apparently gain entrance into stored-product habitats together with their acarid prey, all being introduced therein by birds and small scavengers (Woodroffe, 1953). *Hemisarcoptes* spp., natural enemies of armoured scale insects, are transported during the deutonymphal stage (hypopus) by lady beetles (Coccinellidae) of the genus *Chilocorus*, which seek out and feed on the same diet (Gerson *et al.*, in press). The beetles are unharmed by the deutonymphs, which lack functional mouth parts.
2. Most of these successful natural enemies, while not very specific, show distinct preferences for prey of one group, usually at the order or family level. *Bdellodes* spp. attack mainly collembolans (Wallace and Walters, 1974), *B. tarsalis* prefers stored-product moths (Haines, 1981). *Cheyletus* spp. feed mostly on acarid mites (Hughes, 1976), and *Hemisarcoptes* spp. are restricted to armoured scale insects (Gerson *et al.*, in press). *Macrocheles* spp. feed on nematodes, acarid mites and small fly larvae as well as their eggs, but prefer the latter diet (Axtell

and Rutz, 1986), and *Phytoseiulus persimilis* preys only on webbing spider mites. *A. agilis*, however, will feed on any invertebrate it can capture and puncture (Sorensen et al., 1976). The diet of *Euseius* spp. includes thrips, mites, pollen, and honeydew (McMurtry and Rodriguez, 1987). Some further remarks on specificity are added below.

3. All successful acarine biocontrol agents noted, except *Hemisarcoptes* spp. and *P. persimilis*, can survive in the field in the absence of their preferred diets. Some phytoseiids, listed by McMurtry (1982) among the important phytoseiid predators, may even feed on the leaf tissue.
4. The power of increase (or reproductive potential) of the mites listed in Table 3 is usually similar to or greater than that of the pests. Logan (1982) observed that the intrinsic rate of natural increase, r_m, of phytoseiids is often similar to that of their spider mite prey. *Cheyletus eruditus* develops somewhat faster than its acarid diet under slightly warmer and drier conditions, increasing the chances for successful pest control (Solomon, 1969). Species of *Blattisocius, Macrocheles,* and *Hemisarcoptes* reproduce more rapidly than their prey (Haines, 1981; Axtell and Rutz, 1986; Gerson and Schneider, 1981). On the other hand, *A. agilis* produces only two annual generations in California, fewer than its various prey species (Sorensen et al., 1976).
5. All authors agree that mite predators are rather voracious; the only exception is *Hemisarcoptes coccophagus*, which can complete a generation on an individual armoured scale insect host (Gerson and Schneider, 1981).
6. Although predaceous acarines are usually more affected by pesticides than the pests they feed on, some mites possess a natural lack of sensitivity, or have evolved resistance, to certain of these chemicals (Hoy, 1985; Parent, 1961; Žďárková and Horak, 1987).
7. Acarine biocontrol agents seem to have controlled injurious mites somewhat more frequently than other groups of pests. This inference may be related to more information being available about phytoseiids (of which several species, additional to those listed in Table 3, are known to limit pest mite populations), due to intensive research aimed at this group of predators. Some phytoseiids may, however, also control phytophagous insects such as thrips (Hansen, 1988; Tanigoshi et al., 1984) or even other pests, adding to the uncertainty of this premise. On the other hand, if consistent, the greater suitability of mites for control by predaceous acarines could be related to the smaller size of these target pests (see below).
8. No single type of habitat, whether stable (pastures), intermediately stable (orchards, stored products) or unstable (cow dung pads, greenhouses) seems to be better (or worse) for the biological control

of pests by mites. Hall *et al.* (1980) concluded that intermediately stable habitats were more conducive to success in classical biological control.

The 'portrait' of a potentially successful acarine biocontrol agent which thus emerges is not too different from that formulated for other effective natural enemies. Such an agent has well-developed searching and dispersal abilities, is rather voracious and has a distinct preference for the pest species (and its relatives), though not to the exclusion of other diets (so that it possesses the ability to survive pest scarcity). Its power of increase is at least as high as that of its main prey, and it may be more effective against mite pests. The fact that the members of two successful mite predator genera, *Anystis* and *Hemisarcoptes*, do not quite fit into this mould probably reflects the heterogeneity of acarine biocontrol agents and emphasizes the need for more research.

Some remarks remain to be added about the dispersal of mite predators, their specificity, and the effect of size. As noted, some predaceous mites have surmounted the problem of reaching their discrete prey habitats by being phoretic on insects which may themselves serve as food (*B. tarsalis*, *Macrocheles* spp.) or which feed on the same pest (*Hemisarcoptes* spp.). This points to the need for better understanding of the relationships between the mites and their transporters. Binns (1982) has begun to explore this association by analysing the literature and placing mite phoresy within the context of dispersal. He also emphasized the importance of the phoretic dispersal of mites for the purposes of biological control. Two examples will suffice to demonstrate this point. Bedford (1949) used *Chilocorus* beetles to transport *Hemisarcoptes* sp. to Bermuda during a project to control armoured scale insects on cedar trees there. The need for efficient transporters to bring *Macrocheles glaber* (Muller) onto fresh cow dung pads in order to control the breeding Australian bushfly therein was emphasized by Wallace *et al.* (1979).

Reported observations on the specificity (and non-specificity) of efficient acarine natural enemies suggest that 'specificity' should perhaps be viewed in a more dynamic way. Any of the discussed mite predators could be placed somewhere along a continuum, one extreme of which may accommodate specific feeders like *P. persimilis*, the other general predators like *Anystis* spp. Other mites may be located at different points along this continuum, their precise location at any time being dependent on their ages, ambient humidity, temperature and light conditions (i.e. van Houten *et al.*, 1988), the availability of different diets, the nature of the most recent meal (Congdon and McMurtry, 1988) as well as other factors. It follows that a high degree of specificity cannot always be used to predict the probable success of a given acarine biocontrol agent. More discerning preference tests (Congdon and McMurtry, 1988) may even show that the absolute acarine 'general predator' is a rather rare animal. Strict specificity is an obvious disadvantage during periods of prey scarcity. Predators with such diets are less likely to defend themselves; when *P. persimilis* and *T. occidentalis* occurred together on hop leaves, the former did not attack the eggs and larvae of the latter, which was feeding on the immatures of *P. persimilis*. Pruszynski and Cone (1972) therefore considered *T. occidentalis* to be a better candidate for controlling *T. urticae* on hops.

The significance of sheer size for acarine predators has several times been reported. The large size of the erythraeid *Lasioerythraeus johnstoni* Welbourn and Young in relation to its host, the tarnished plant bug, *Lygus lineolaris* (Palisot de Beauvois), was considered (Young and Welbourn, 1987) to be a promising character for pest control by this mite. Axtell and Rutz (1986) noted that the mesostigmatid *Fuscuropoda vegetans* feeds on first instar fly larvae but cannot subdue larger and more vigorous second and third instars. Bakker and Sabelis (1987) reported that first as well as second instar greenhouse thrips, *Thrips tabaci*, defended themselves from attack by the phytoseiid *Amblyseius barkeri* (Hughes) (=*Amblyseius mckenziei* Schuster and Pritchard) by wagging. The defensive behaviour of the older nymphs was more vigorous and frequent. Very small acarine biocontrol agents, such as the Pyemotidae, can subdue much larger prey by injecting them with a quick-acting paralysing toxin (Moser, 1975). An injected toxin was also postulated to cause death of cockroaches attacked by *Pimeliaphilus cunliffei* Jack (Field *et al.*, 1966). The 'counter-attack' success of male and female *Schizotetranychus celarius* (Banks) against their phytoseiid predator was highest against the latter's larvae (Saito, 1986). A prey-size to predator-size analysis (Hespenheide, 1973) could be instructive in this context as more information becomes available.

Data in Table 3 suggest that some predatory mites do better when controlling aggregated pest populations, or those within a restricted or closed habitat. The more efficient phytoseiids succeed in controlling colonial, webbing spider mites (Sabelis, 1985), usually in greenhouses. *H. coccophagus* does best when attacking dense armoured scale insect colonies (Gerson and Schneider, 1981). *Macrocheles glaber* may control the Australian bushfly (*Musca vetustissima* Walker) within dung pads (Wallace *et al.*, 1979), and *C. eruditus* reduces acarid mite infestations within stored products (Žďárková, 1986). This suggests that many acarine biocontrol agents are quite sensitive to prey density, and possibly do best when pest populations are strongly clumped. This proposition is supported by observations that various predaceous mites, which are not particularly effective in the field, seriously damage insect pests in laboratory cultures. These natural enemies, which include the prostigmatid *Saniosulus nudus* Summers and *Acaropsis* sp. (Gerson and Blumberg, 1969; Strong *et al.*, 1959), must consequently be controlled by chemical means. However, work with *T. pyri* showed that the distribution of this efficient predator is random with respect to that of the pest prey it controls, namely *P. ulmi* (Nyrop, 1988). It was concluded that the non-aggregation of this successful biological control agent possibly stemmed from a search strategy evolved to cope with non-aggregated prey. This point, which is of general interest, requires further observations.

Harris (1973) and Goeden (1983) proposed scoring systems for predicting the efficacy of arthropods introduced for weed control shortly before, during and after their initial release. As more data become available about acarine biocontrol agents, a similar system may perhaps be devised for their evaluation.

42
References

Abdel-Malek, A. (1948) The biology of *Aedes trivittatus*. *J. Econ. Entomol.*, **41**, 951–4.

Abo Elghar, M.R., Elbadry, E.L., Hassan, S.M. and Kilany, S.M. (1969) Studies on the feeding, reproduction and development of *Agistemus exsertus* on various pollen species (Acarina: Stigmaeidae). *Z. Angew. Entomol.*, **63**, 282–4.

Aeschlimann, J.P. and Vitou, J. (1986) Observations on the association of *Allothrombium* sp. (Acari: Thrombidiidae) mites with lucerne aphid populations in the Mediterranean region. In *Ecology of Aphidophaga* (ed. I. Hodek), Academia, Prague, pp. 405–10.

Anderson, R.M. (1978) The regulation of host population growth by parasitic species. *Parasitology*, **76**, 119–57.

Asano, S. and Kamei, M. (1982) Ovicidal activity of cycloprate for several phytophagous mite species and its relationship with the test host plant. *Appl. Entomol. Zool.*, **17**, 67–74.

Avidov, Z., Blumberg, D. and Gerson, U. (1968) *Cheletogenes ornatus* (Acarina: Cheyletidae), a predator of the chaff scale on citrus in Israel. *Isr. J. Entomol.*, **3**, 77–94.

Axtell, R.C. (1963) Effect of Macrochelidae (Acarina: Mesostigmata) on house fly production from dairy cattle manure. *J. Econ. Entomol.*, **56**, 317–21.

Axtell, R.C. (1966) Comparative toxicities of insecticides to house fly larvae and *Macrocheles muscaedomesticae*, a mite predator of the house fly. *J. Econ. Entomol.*, **59**, 1128–30.

Axtell, R.C. and Edwards, T.D. (1983) Efficacy and nontarget effects of Larvadex® as a feed additive for controlling house flies in caged-layer poultry manure. *Poultry Sci.*, **62**, 2371–7.

Axtell, R.C. and Rutz, D.A. (1986) Role of parasites and predators as biological fly control agents in poultry production facilities. *Misc. Publ. Entomol. Soc. Am.*, **61**, 88–100.

Bakker, F.M. and Sabelis, M.W. (1987) Attack success of *Amblyseius mckenziei* and the stage related defensive capacity of thrips larvae. In *Working Group 'Integrated Control in Glasshouses'* (eds B. Nedstam, L. Stengard Hansen and J.C. van Lenteren), EPRS/WPRS, Budapest, pp. 26–9.

Ball, J.C. (1982) Impact of fungicides and miticides on predatory and phytophagous mites associated with pecan foliage. *Environ. Entomol.*, **11**, 1001–4.

Barker, P.S. (1968) Effect of food quality on the reproduction of Mesostigmata: a review. *Manitoba Entomol.*, **2**, 46–8.

Bartlett, B.R. (1964) The toxicity of some pesticide residues to adult *Amblyseius hibisci*, with a compilation of the effects of pesticides upon phytoseiid mites. *J. Econ. Entomol.*, **57**, 559–63.

Bedford, E.C.G. (1949) Report of the plant pathologist. In *Report of the Department*

REFERENCES

of Agriculture for the year 1949. Bermuda Board of Agriculture, pp. 11–19.

Behan-Pelletier, V.M. and Hill, S.B. (1983) Feeding habits of sixteen species of Oribatei (Acari) from an acid peat bog, Glenamoy, Ireland. *Rev. Ecol. Biol. Sol.*, **20**, 221–67.

Bellows, T.S. Jr and Morse, J.G. (1988) Residual toxicity following dilute or low-volume applications of insecticides used for control of California red scale (Homoptera: Diaspididae) to four beneficial species in a citrus agroecosystem. *J. Econ. Entomol.*, **81**, 892–8.

Bergman, J.M. and Tingey, W.M. (1979) Aspects of interaction between plant genotypes and biological control. *Bull. Entomol. Soc. Am.*, **25**, 275–9.

Binns, E.S. (1982) Phoresy as migration – some functional aspects of phoresy in mites. *Biol. Rev.*, **57**, 571–620.

Boethel, D.J. and Eikenbary, R.D. (eds) (1986) *Interactions of Plant Resistance and Parasitoids and Predators of Insects*. Ellis Horwood, Chichester, 224 pp.

Böttger, K. (1970) Die Ernährungsweise der Wassermilben (Hydrachnellae, Acari). *Int. Rev. ges. Hydrobiol.*, **55**, 895–912.

Brickhill, C.D. (1958) Biological studies of two species of tydeid mites from California. *Hilgardia*, **27**, 601–20.

Bruce, W.A. (1983) Mites as biological control agents of stored product pests. In *Biological Control of Pests by Mites* (eds M.A. Hoy, G.L. Cunningham and L. Knutson), University of California Special Publication no. 3304, pp. 74–8.

Bruce, W.A. (1989) Artificial diet for the parasitic mite *Pyemotes tritici* (Acari: Pyemotidae). *Exp. Appl. Acarol.*, **6**, 11–18.

Chiang, H.C. (1970) Effects of manure applications and mite predation on corn rootworm populations in Minnesota. *J. Econ. Entomol.*, **63**, 934–6.

Clancy, D.W. and Pollard, H.N. (1952) The effect of DDT on mite and predator populations in apple orchards. *J. Econ. Entomol.*, **45**, 109–14.

Collyer, E. (1964a) Phytophagous mites and their predators in New Zealand orchards. *NZ J. Agric. Res.*, **7**, 551–68.

Collyer, E. (1964b) A summary of experiments to demonstrate the role of *Typhlodromus pyri* Scheut. in the control of *Panonychus ulmi* (Koch) in England. *Acarologia*, **6**, (h.s.), 363–71.

Congdon, B.D. and McMurtry, J.A. (1988) Prey selectivity in *Euseius tularensis* (Acari: Phytoseiidae). *Entomophaga*, **33**, 281–7.

Congdon, B.D. and Tanigoshi, L.K. (1983) Indirect toxicity of dimethoate to the predaceous mite *Euseius hibisci* (Chant) (Acari: Phytoseiidae). *Environ. Entomol.*, **12**, 933–5.

Cook, D.R. (1974) Water mite genera and subgenera. *Mem. Am. Entomol. Soc. Wash.*, **21**, 1–80.

Croft, B.A. and Blyth, E.J. (1979) Aspects of the functional, ovipositional and starvation response of *Amblyseius fallacis* to prey density. In *Recent Advances in Acarology* (ed. J.G. Rodriguez), Academic Press, New York, Vol. 1, pp. 41–7.

Croft, B.A. and Brown, A.W.A. (1975) Responses of arthropod natural enemies to insecticides. *Annu. Rev. Entomol.*, **20**, 285–335.

Croft, B.A. and Nelson, E.E. (1972) An index to predict efficient interactions of *Typhlodromus occidentalis* in control of *Tetranychus mcdanieli* in southern California apple trees. *J. Econ. Entomol.*, **65**, 310–12.

Crofton, H.D. (1971) A quantitative approach to parasitism. *Parasitology*, **62**, 179–93.

Currie, G.A. (1934) The bdellid mite *Biscirus lapidarius* Kramer, predatory on the lucerne flea *Sminthurus viridis* L. in Western Australia. *J. Aust. Counc. Sci. Ind. Res.*, **7**, 9–20.

Daneshvar, H. and Rodriguez, J.C. (1975) Toxicity of organophosphorus systemic pesticides to predator mites and prey. *Entomol. Exp. Appl.*, **18**, 297–301.

Davids, C. and Schoots, C.J. (1975) The influence of the water mite species *Hydrachna conjecta* and *H. cruenta* (Acari, Hydrachnellae) on the egg production of the Corixidae *Sigara striata* and *Cymatia coleoptrata* (Hemiptera). *Verh. Int. Verein. Limnol.*, **19**, 3079–82.

Davies, D.M. (1959) The parasitism of black flies (Diptera, Simuliidae) by larval water mites mainly of the genus *Sperchon*. *Can. J. Zool.*, **73**, 353–69.

Dean, R.W., Palmiter, D.H. and Hickey, K.D. (1966) Suppression of European red mite by mildew fungicide programs. *J. Econ. Entomol.*, **59**, 742.

Dicke, M. and Sabelis, M.W. (1988) How plants obtain predatory mites as bodyguards. *Neth. J. Zool.*, **38**, 148–65.

Dicke, M., Sabelis, M.W. and Groeneveld, A. (1986) Vitamin A deficiency modifies response of predatory mite *Amblyseius potentillae* to volatile kairomone of two-spotted spider mite. *J. Chem. Ecol.*, **12**, 1389–96.

Dover, M.J., Croft, B.A., Welch, S.M. and Tummalo, R.L. (1979) Biological control of *Panonychus ulmi* (Acarina: Tetranychidae) by *Amblyseius fallacis* (Acarina: Phytoseiidae) on apple: a prey–predator model. *Environ. Entomol.*, **8**, 282–92.

Downing, R.S. and Moilliet, T.K. (1967) Relative densities of predacious and phytophagous mites on three varieties of apple tree. *Can. Entomol.*, **99**, 738–41.

Downing, R.S. and Moilliet, T.K. (1972) Replacement of *Typhlodromus occidentalis* by *T. caudiglans* and *T. pyri* (Acarina: Phytoseiidae) after cessation of sprays on apple trees. *Can. Entomol.*, **104**, 937–40.

Drummond, F.A., Logan, P.A., Casagrande, R.A. and Gregson, F.A. (1985) Host specificity tests of *Chrysomelobia labidomerae*, a mite parasitic on the Colorado potato beetle. *Int. J. Acarol.*, **11**, 169–72.

Eickwort, G.C. (1983) Potential use of mites as biological control agents of leaf-feeding insects. In *Biological Control of Pests by Mites* (eds M.A. Hoy, G.L. Cunningham and L. Knutson), University of California Special Publication no. 3304, pp. 41–52.

Field, G., Savage, L.B. and Duplessis, R.J. (1966) Note on the cockroach mite, *Pimeliaphilus cunliffei* (Acarina: Pterygosomidae) infesting oriental, German and American cockroaches. *J. Econ. Entomol.*, **59**, 1532.

Fleschner, C.A. (1958) Field approach to population studies of tetranychid mites on citrus and avocado in California. *Proc. 10th Int. Congr. Entomol.*, Montreal, 17–25 August 1956, Vol. 2, pp. 669–74.

Garman P. (1948) Mite species from apple trees in Connecticut. *Bull. Conn. Agric. Exp. Stn*, New Haven, No. 520, 27 pp.

Gauchat, C.A. (1972) A note on *Podapolipoides grassi* Berlese (Acarina: Podapolipidae), a parasite of *Chortoicetes terminifera* Walker, the Australian plague locust. *J. Aust. Entomol. Soc.*, **11**, 259.

REFERENCES

Gerson, U. (1967) Observations on *Hemisarcoptes coccophagus* Meyer (Astigmata: Hemisarcoptidae), with a new synonym. *Acarologia*, **9**, 632–8.

Gerson, U. (1985) Webbing. In *Spider Mites: Their Biology, Natural Enemies and Control*, (eds W. Helle, and M.W. Sabelis), Elsevier, Amsterdam, vol. 1A, pp. 223–32.

Gerson, U. and Aronowitz, A. (1980) Feeding of the carmine spider mite on seven host plant species. *Entomol. Exp. Appl.*, **28**, 109–15.

Gerson, U. and Aronowitz, A. (1981) Spider mite webbing. V. The effect of various host plants. *Acarologia*, **22**, 277–81.

Gerson, U. and Blumberg, D. (1969) Biological notes on the mite *Saniosulus nudus*. *J. Econ. Entomol.*, **62**, 729–30.

Gerson, U. and Cohen, E. (1989) Resurgences of spider mites (Acari: Tetranychidae) induced by synthetic pyrethroids. *Exp. Appl. Acarol.*, **6**, 29–46.

Gerson, U. and Schneider, R. (1981) Laboratory and field studies on the mite *Hemisarcoptes coccophagus* Meyer (Astigmata: Hemisarcoptidae), a natural enemy of armored scale insects. *Acarologia*, **22**, 199–208.

Gerson, U., O Connor, B.M. and Houck, M.A. Acari. In *Armored Scale Insects, Biology, Natural Enemies and Control* (ed. D. Rosen), Elsevier, Amsterdam, in press.

Gilliatt, F.C. (1935) Some predators of the European red mite, *Paratetranychus pilosus* C. & F., in Nova Scotia. *Can. J. Res.*, **13D**, 19–38.

Gledhill, T., Cowley, J. and Gunn, R.J.M. (1982) Some aspects of the host:parasite relationships between adult blackflies (Diptera; Simuliidae) and larvae of the water-mite *Sperchon setiger* (Acari; Hydrachnellae) in a small chalk stream in southern England. *Freshw. Biol.*, **12**, 345–57.

Glendenning, R. (1931) The progress of parasite introduction in British Columbia. *Proc. Entomol. Soc. Br. Colum.*, **28**, 29–32.

Goeden, R.D. (1983) Critique and revision of Harris' scoring system for selection of insect agents in biological control of weeds. *Prot. Ecol.*, **5**, 287–301.

Grafton-Cardwell, E.E. and Hoy, M.A. (1983) Comparative toxicity of avermectin B_1 to the predator *Metaseiulus occidentalis* (Nesbitt) (Acari: Phytoseiidae) and the spider mites (*Tetranychus urticae* Koch and *Panonychus ulmi* (Koch) (Acari: Tetranychidae). *J. Econ. Entomol.*, **76**, 1216–20.

Graham, W.M. (1970) Warehouse ecology studies of bagged maize in Kenya – II. Ecological observations of an infestation by *Ephestia (Cadra) cautella* (Walker) (Lepidoptera, Phycitidae). *J. Stored Prod. Res*, **6**, 157–67.

Griffiths, D.A. (1960) Some field habitats of mites of stored food products. *Ann. Appl. Biol.*, **48**, 134–44.

Hagley, E.A.C. and Simpson, C.M. (1983) Effects of insecticides on predators of the pear psylla, *Psylla pyricola* (Hemiptera: Psyllidae), in Ontario. *Can. Entomol.*, **115**, 1409–14.

Haines, C.P. (1981) Laboratory studies on the role of an egg predator, *Blattisocius tarsalis* (Berlese) (Acari: Ascidae), in relation to the natural control of *Ephestia cautella* (Walker) (Lepidoptera: Pyralidae) in warehouses. *Bull. Entomol. Res.*, **71**, 555–74.

Hall, R.W., Ehler, L.E. and Bisabri-Ershadi, B. (1980) Rate of success in classical biological control of arthropods. *Bull. Entomol. Soc. Am.*, **26**, 111–14.

Hamai, J. and Huffaker, C.B. (1978) Potential of predation by *Metaseiulus occidentalis* in compensating for increased, nutritionally induced, power of increase of *Tetranychus urticae*. *Entomophaga*, **23**, 225–37.

Haney, P.B., Luck, R.F. and Moreno, D.S. (1987) Increases in densities of the citrus red mite, *Panonychus citri* (Acarina: Tetranychidae), in association with the Argentine ant, *Iridomyrmex humilis* (Hymenoptera: Formicidae), in southern California citrus. *Entomophaga*, **32**, 49–57.

Hansen, L.S. (1988) Control of *Thrips tabaci* (Thysanoptera: Thripidae) on glasshouse cucumber using large introductions of predatory mites *Amblyseius barkeri* (Acarina: Phytoseiidae). *Entomophaga*, **33**, 33–42.

Harris, P. (1973) The selection of effective agents for the biological control of weeds. *Can. Entomol.*, **105**, 1495–503.

Hassan, S.A., Bigler, F., Bogenschutz, H., Boller, E., Brun, J. *et al.* (1988) Results of the fourth joint pesticide testing programme carried out by the IOBC/WPRS-Working Group 'Pesticides and Beneficial Arthropods'. *J. Appl. Entomol.*, **105**, 321–9.

Herne, D.H.C. and Putman, W.L. (1966) Toxicity of some pesticides to predacious arthropods in Ontario peach orchards. *Can. Entomol.*, **98**, 936–42.

Herren, H.R., Bird, T.J. and Nadel, D.J. (1987) Technology for automated aerial releases of natural enemies of the cassava mealybug and cassava green mite. *Insect Sci. Appl.*, **8**, 883–5.

Hespenheide, H.A. (1973) Ecological inferences from morphological data. *Annu. Rev. Ecol. Syst.*, **4**, 213–29.

Hill, R.L. and Stone, C. (1985) Spider mites as control agents for weeds. In *Spider Mites: Their Biology, Natural Enemies and Control* (eds W. Helle and M.W. Sabelis), Elsevier, Amsterdam, vol. 1B, pp. 443–8.

Hislop, R.G., Auditore, P.J., Weeks, B.L. and Prokopy, R.J. (1981) Repellency of pesticides to the mite predator *Amblyseius fallacis*. *Prot. Ecol.*, **3**, 253–7.

Hokkonen, H.M.T. (1985) Success in classical biological control. *CRC Crit. Rev. Plant Sci.*, **3**, 35–72.

Howard, L.O. (1930) *A History of Applied Entomology (Somewhat Anecdotal)*. Smithsonian Misc. Coll., Publ. no. 3065, 564 pp.

Hoy, M.A. (1982) Genetics and genetic improvement of the Phytoseiidae. In *Recent Advances in Knowledge of the Phytoseiidae* (ed. M.A. Hoy), University of California Division of Agricultural Science Publication no. 3284, pp. 72–89.

Hoy, M.A. (1985) Recent advances in genetics and genetic improvement of the Phytoseiidae. *Annu. Rev. Entomol.*, **30**, 345–70.

Hoy, M.A., Castro, D. and Cahn, D. (1982) Two methods for large scale production of pesticide-resistant strains of the spider mite predator *Metaseiulus occidentalis* (Nesbitt) (Acarina: Phytoseiidae). *Z. Angew. Entomol.*, **94**, 1–9.

Hoy, M.A., Cunningham, G.L. and Knutson, L. (eds) (1983) *Biological Control of Pests by Mites*. University of California Special Publication no. 3304, 185 pp.

Huffaker, C.B., Simmonds, F.J. and Laing, J.E. (1976) The theoretical and empirical basis of biological control. In *The Theory and Practice of Biological Control* (eds C.B. Huffaker and P.S. Messenger), Academic Press, New York, pp. 41–78.

REFERENCES

Huffaker, C.B., van de Vrie, M. and McMurtry, J.A. (1969) The ecology of tetranychid mites and their natural enemies. *Annu. Rev. Entomol.*, **14**, 125–74.

Huggans, J.L. and Blickenstaff, C.C. (1966) Parasites and predators of grasshoppers in Missouri. *Miss. Agric. Exp. Stn, Res. Bull.* 903, p. 40.

Hughes, A.M. (1976) *The Mites of Stored Food and Houses*, Her Majesty's Stationery Office, London, 400 pp.

Hunter, W.D. and Hinds, W.E. (1904) The Mexican cotton boll weevil. *USDA Div. Entomol.*, Bull. No. 45, 116 pp.

Husband, R.W. (1974) Lectotype designation for *Locustacarus trachealis* Ewing and a new species of *Locustacarus* (Acarina: Podapolipidae) from New Zealand. *Proc. Entomol. Soc. Wash.*, **76**, 52–9.

Imbriani, J.L. and Mankau, R. (1983) Studies on *Lasioseius scapulatus*, a mesostigmatid mite predaceous on nematodes. *J. Nematol.*, **15**, 523–8.

Karg, W. (1983) Verbreitung und Bedeutung von Raubmilben der Cohors Gamasina als Antagonisten von Nematoden. *Pedobiologia*, **25**, 419–32.

Knop, N.F. and Hoy, M.A. (1983) Tydeid mites in vineyards. *Calif. Agric.*, **37**, 16–18.

Kovalev, O.V. (1973) Modern outlooks of biological control of weed plants in the U.S.S.R. and the international phytophagous exchange. In *Proc. 2nd Int. Symp. Biol. Contr. Weeds* (ed. P.H. Dunn), Commonwealth Agricultural Bureaux, pp. 166–72.

Lagerlof, J. and Andren, O. (1988) Abundance and activity of soil mites (Acari) in four cropping systems. *Pedobiologia*, **32**, 129–45.

Laing, J.E. (1973) Evaluating the effectiveness of *Paracheyletia bakeri* (Acarina: Cheyletidae) as a predator of the two-spotted spider mite, *Tetranychus urticae*. *Ann. Entomol. Soc. Am.*, **66**, 641–6.

Laing, J.E. and Knop, N.F. (1983) Potential use of predaceous mites other than Phytoseiidae for biological control of orchard pests. In *Biological Control of Pests by Mites* (eds M.A. Hoy, G.L. Cunningham and L. Knutson), University of California Special Publication no. 3304, pp. 28–35.

Laing, J.E. and Osborn, J.A.L. (1974) The effect of prey density on the functional and numerical responses of three species of predatory mites. *Acarologia*, **19**, 267–77.

Lanciani, C.A. (1979a) The influence of parasitic water mites on the instantaneous death rate of their hosts. *Oecologia*, **44**, 60–2.

Lanciani, C.A. (1979b) Water mite-induced mortality in a natural population of the mosquito *Anopheles crucians* (Diptera: Culicidae). *J. Med. Entomol.*, **15**, 529–32.

Lanciani, C.A. (1982) Parasite-induced reductions in the survival and reproduction of the backswimmer *Buenoa scimitra* (Hemiptera: Notonectidae). *Parasitology*, **85**, 593–603.

Lanciani, C.A. (1983) Overview of the effects of the water mite parasitism on aquatic insects. In *Biological Control of Pests by Mites* (eds M.A. Hoy, G.L. Cunningham and L. Knutson), University of California Special Publication no. 3304, pp. 86–90.

Lanciani, C.A. (1986) Reduced survivorship in *Dasyhelea mutabilis* (Diptera: Ceratopogonidae) parasitized by the water mite *Tyrrellia circularis* (Acariformes: Limnesiidae). *J. Parasitol.*, **72**, 613–14.

Lanciani, C.A. (1988) Defensive consumption of parasitic mites by *Anopheles crucians* larvae. *J. Am. Mosquito Control Assoc.*, **4**, 195.

Lanciani, C.A. and Boyett, J.M. (1980) Demonstrating parasitic water mite-induced mortality in natural host populations. *Parasitology*, **81**, 465–75.

Lanciani, C.A. and Boyt, A.D. (1977) The effect of a parasitic water mite, *Arrenurus pseudotenuicollis* (Acari: Hydrachnellae), on the survival and reproduction of the mosquito *Anopheles crucians* (Diptera: Culicidae), *J. Med. Entomol.*, **14**, 10–15.

Lange, A.B., Drozdovskii, E.M. and Bushkovskaya, L.M. (1974a) The anystis mite – an effective predator of small phytophages. *Zashchita Rast.*, **1974**, 26–8 (in Russian).

Lange, A.B., Drozdovskii, E.M. and Bushkovskaya, L.M. (1974b) Collecting and releasing anystids. *Zashchita Rast.*, **1974**, 33–4 (in Russian).

Lawrence, R.F. (1940) A note on some mite parasites of *Locusta migratoria migratorioides* R. & F. *J. Entomol. Soc. S. Afr.*, **3**, 173–8.

Lehmann, H.D. (1970) Zur Wirkung von Neguvon® auf Milben der Familie Pterygosomidae. *Salamandra*, **6**, 128–30.

Lindquist, R.K. and Wolgamott, M.L. (1980) Toxicity of acephate to *Phytoseiulus persimilis* and *Tetranychus urticae*. *Environ. Entomol.*, **9**, 389–92.

Lister, A., Block, W. and Usher, M.B. (1988) Arthropod predation in an Antarctic terrestrial community. *J. Anim. Ecol.*, **57**, 957–71.

Lister, A., Usher, M.B. and Block, W. (1987) Description and quantification of field attack rates by predatory mites: an example using an electrophoretic method with a species of Antarctic mite. *Oecologia*, **72**, 185–91.

Lockwood, J.A., Sparks, T.C. and Story, R.N. (1984) Evolution of insect resistance to insecticides: a reevaluation of the roles of physiology and behavior. *Bull. Entomol. Soc. Am.*, **30**, 41–51.

Logan, J.A. (1982) Recent advances and new directions in phytoseiid population models. In *Recent Advances in Knowledge of the Phytoseiidae* (ed. M.A. Hoy), University of California Division of Agricultural Science Publication no. 3284, pp. 49–71.

Lord, F.T. (1947) The influence of spray programs on the fauna of apple orchards in Nova Scotia. II. Oystershell scale. *Can. Entomol.*, **79**, 196–209.

Lord, F.T. (1949) The influence of spray programs on the fauna of apple orchards in Nova Scotia. III. Mites and their predators. *Can. Entomol.*, **81**, 202–14, 217–30.

Lord, F.T. and MacPhee, A.W. (1953) The influence of spray programs on the fauna of apple orchards in Nova Scotia. VI. Low temperatures and the natural control of the oystershell scale, *Lepidosaphes ulmi* (L.) (Homoptera: Coccidae). *Can. Entomol.*, **85**, 282–91.

Luck, R.F., Shepard, B.M. and Kenmore, P.E. (1988) Experimental methods for evaluating arthropod natural enemies. *Annu. Rev. Entomol.*, **33**, 367–91.

MacPhee, A.W. and Sanford, K.H. (1954) The influence of spray programs on the fauna of apple orchards in Nova Scotia. VII. Effects on some beneficial arthropods. *Can Entomol.*, **86**, 128–35.

Mallea, A.R., Macola, G.S., Garcia Saez, J.G. and Lamati, S.J. (1983) Localizacion de *Pyemotes* sp. (Acarina – Pyemotidae) sobre *Saissetia oleae* Bern. (Homoptera – Lecaniidae). *Intersectum*, **15**, 8–10.

REFERENCES

Mankau, R. (1980) Biological control of nematode pests by natural enemies. *Annu. Rev. Phytopathol.*, **18**, 415–40.

Mansour, F., Ascher, K.R.S. and Omari, N. (1987) Effects of neem (*Azadirachta indica*) seed kernel extracts from different solvents on the predaceous mite *Phytoseiulus persimilis* and the phytophagous mite *Tetranychus cinnabarinus*. *Phytoparasitica*, **15**, 125–30.

Markwick, N.P. (1986) Detecting variability and selecting for pesticide resistance in two species of phytoseiid mites. *Entomophaga*, **31**, 225–36.

McMurtry, J.A. (1982) The use of phytoseiids for biological control: progress and future prospects. In *Recent Advance in Knowledge of the Phytoseiidae* (ed. M.A. Hoy), University of California Division of Agricultural Science Publication no. 3284, p. 23–48.

McMurtry, J.A. and Rodriguez, J.G. (1987) Nutritional ecology of phytoseiid mites. In *Nutritional Ecology of Insects, Mites and Spiders* (eds F. Slansky Jr and J.G. Rodriguez), Wiley, Chichester, pp. 609–44.

Mendel, Z. and Gerson, U. (1982) Is the mite *Lorryia formosa* Cooreman (Prostigmata: Tydeidae) a sanitizing agent in citrus groves? *Oecol. Appl.*, **3**, 47–51.

Meyer, M.K.P. (Smith) and Ueckermann, E.A. (1987) A taxonomic study of some Anystidae (Acari: Prostigmata). *Entomol. Mem. Dep. Agric. Wat. Sup. Repub. S. Afr.*, **68**, 1–37.

Miller, D.R., and Kosztarab, M. (1979) Recent advances in the study of scale insects. *Annu. Rev. Entomol.*, **24**, 1–27.

Moraes, G.J. de and McMurtry, J.A. (1987) Physiological effect of the host plant on the suitability of *Tetranychus urticae* as prey for *Phytoseiulus persimilis* (Acari: Tetranychidae, Phytoseiidae). *Entomophaga*, **32**, 35–8.

Moser, J.C. (1975) Biosystematics of the straw itch mite with special reference to nomenclature and dermatology. *Trans. R. Entomol. Soc. London*, **127**, 185–91.

Motoyama, N., Rock, G.C. and Dauterman, W.C. (1970) Organophosphorus resistance in an apple orchard population of *Typhlodromus* (*Amblyseius*) *fallacis*. *J. Econ. Entomol.*, **63**, 1439–42.

Mullen, G.R. (1975) Acarine parasites of mosquitoes. 1. A critical review of all known records of mosquitoes parasitized by mites. *J. Med. Entomol.*, **12**, 27–36.

Muma, M.H. (1975) Mites associated with citrus in Florida. *Univ. Florida Agric. Exp. Stn Bull.* 640A.

Muma, M.H. and Selhime, A.G. (1971) *Agistemus floridanus* (Acarina: Stigmaeidae), a predatory mite, on Florida citrus. *Fla. Entomol.*, **54**, 249–58.

Muraoka, M. and Ishibashi, N. (1976) Nematode-feeding mites and their feeding behaviour. *Appl. Entomol. Zool.*, **11**, 1–7.

Murray, R.A. and Solomon, M.G. (1978) A rapid technique for analysing diets of invertebrate predators by electrophoresis. *Ann. Appl. Biol.*, **90**, 7–10.

Nelson, E.E., Croft, B.A., Howett, A.J. and Jones, A.L. (1973) Toxicity of apple orchard pesticides to *Agistemus fleschneri*. *Environ. Entomol.*, **2**, 219–22.

Newell, I.M. (1979) *Acarus tinctorius* Linnaeus 1767 (Trombidiidae). In *Recent Advances in Acarology* (ed. J.G. Rodriguez), Vol. 2, Academic Press, New York, pp. 425–8.

Nyrop, J.P. (1988) Spatial dynamics of an acarine predator–prey system:

Typhlodromus pyri (Acari: Phytoseiidae) preying on *Panonychus ulmi* (Acari: Tetranychidae). *Environ. Entomol.*, **17**, 1019–31.

Ochieng, R.S., Oloo, G.W. and Amboga, E.O. (1987) An artificial diet for rearing the phytoseiid mite, *Amblyseius teke* Pritchard and Baker. *Exp. Appl. Acarol.*, **3**, 169–73.

Oomen, P.A. (1982) Studies on population dynamics of the scarlet mite, *Brevipalpus phoenicis*, a pest of tea in Indonesia. *Med. Landbouw. Wageningen*, **82–1**, 1–88.

Osborne, L.S. and Petitt, F.L. (1985) Insecticidal soap and the predatory mite, *Phytoseiulus persimilis* (Acari: Phytoseiidae), used in management of the twospotted spider mite (Acari: Tetranychidae) on greenhouse grown foliage plants. *J. Econ. Entomol.*, **78**, 687–91.

Overmeer, W.P.J. (1985) Rearing and handling. In *Spider Mites: Their Biology, Natural Enemies and Control* (eds W. Helle and M.W. Sabelis), Elsevier, Amsterdam, vol. 1B, pp. 161–70.

Overmeer, W.P.J. and van Zon, A.Q. (1983) The effect of different kinds of food on the induction of diapause in the predacious mite *Amblyseius potentillae*. *Entomol. Exp. Appl.*, **33**, 27–30.

Overmeer, W.P.J. and van Zon, A.Q. (1984) The preference of *Amblyseius potentillae* (Garman) (Acarina: Phytoseiidae) for certain plant substrates. In *Acarology VI* (eds D.A. Griffiths and C.E. Bowman), Ellis Horwood, Chichester, Vol. 1, pp. 591–6.

Parent, B. (1961) Effets de certains produits antiparasitaires sur *Typhlodromus rhenanus* (Oudms) et *Mediolata mali* (Ewing), deux acariens prédateurs du Tétranyque rouge du pommier. *Ann. Entomol. Soc. Quebec*, **6**, 55–61.

Pemberton, R.W. and Turner, C.E. (1989) Occurrence of predatory and fungivorous mites in leaf domatia. *Am. J. Bot.*, **76**, 105–12.

Penman, D.R., Chapman, R.B. and Jesson, K.E. (1981) Effects of fenvalerate and azinphosmethyl on two-spotted spider mite and phytoseiid mites. *Entomol. Exp. Appl.*, **30**, 91–7.

Pfeiffer, D.G. (1986) Effects of field applications of paraquat on densities of *Panonychus ulmi* (Koch) and *Neoseiulus fallacis* (Garman). *J. Agric. Entomol.*, **3**, 322–5.

Pickett, C.H., Gilstrap, F.E., Morrison, R.K. and Bouse, L.F. (1987) Release of predatory mites (Acari: Phytoseiidae) by aircraft for the biological control of spider mites (Acari: Tetranychidae) infesting corn. *J. Econ. Entomol.*, **80**, 906–10.

Price, P.W., Bouton, C.E., Gross, P., McPheron, B.A., Thompson, J.N. and Weis, A.E. (1980) Interactions among three trophic levels: influence of plants on interactions between insect herbivores and natural enemies. *Annu. Rev. Ecol. Syst.*, **11**, 41–65.

Pruszynski, S. and Cone, W.W. (1972) Relationships between *Phytoseiulus persimilis* and other enemies of the twospotted spider mite on hops. *Environ. Entomol.*, **1**, 431–3.

Pulpan, J. and Verner, P.H. (1965) Control of tyroglyphoid mites in stored grain by the predatory mite *Cheyletus eruditus* (Schrank). *Can. J. Zool.*, **43**, 417–32.

Putman, W.L. and Herne, H.C. (1964) Relations between *Typhlodromus caudiglans*

REFERENCES

Schuster (Acarina: Phytoseiidae) and phytophagous mites in Ontario peach orchards. *Can. Entomol.*, **96**, 925–43.

Raffa, K.F. and Priester, T.M. (1985) Synergists as research tools and control agents in agriculture. *J. Agric. Entomol.*, **2**, 27–45.

Ragusa, S. and Swirski, E. (1977) Feeding habits, post-embryonic and adult survival, mating, virility and fecundity of the predaceous mite *Amblyseius swirskii* (Acarina: Phytoseiidae) on some coccids and mealybugs. *Entomophaga*, **22**, 383–92.

Ramakers, P.M.J. (1983) Mass production and introduction of *Amblyseius mckenziei* and *A. cucumeris*. *Bull. SROP/WPRS*, **6**, 203–6.

Rangel, A.F. (1901) Cuarto informe acerca del picudo del algodon (*Insanthonomus grandis*, I.C.Cu.). *Bol. Comis. Parasitol. Agri.*, **1**, 245–61.

Rasmy, A.H., Elbagoury, M.E. and Reda, A.S. (1987) A new diet for reproduction of two predaceous mites *Amblyseius gossipi* and *Agistemus exsertus* (Acari: Phytoseiidae, Stigmaeidae). *Entomophaga*, **32**, 277–80.

Richards, A.M. (1962) The oyster-shell scale *Quadraspidiotus ostreaeformis* (Curtis) in the Christchurch district of New Zealand. *NZ J. Agric. Res.*, **5**, 95–100.

Ripper, W.E. (1956) Effect of pesticides on balance of arthropod populations. *Annu. Rev. Entomol.*, **1**, 403–38.

Ristich, S.S. (1956) Toxicity of pesticides to *Typhlodromus fallacis* (Gar.). *J. Econ. Entomol.*, **49**, 511–15.

Rock, G.C. and Yeargan, D.R. (1973) Toxicity of apple orchard herbicides and growth-regulating chemicals to *Neoseiulus fallacis* and twospotted spider mite. *J. Econ. Entomol.*, **66**, 1342–3.

Rockett, C.L. (1980) Nematode predation by oribatid mites (Acari: Oribatida). *Int. J. Acarol.*, **6**, 219–24.

Rodriguez, J.G., Wade, C.F. and Wells, C.N. (1962) Nematodes as a natural food for *Macrocheles muscaedomesticae* (Acarina: Macrochelidae), a predator of the house fly egg. *Ann. Entomol. Soc. Am.*, **55**, 507–11.

Rosen, D. and Huffaker, C.B. (1983) An overview of desired attributes of effective biological control agents, with particular emphasis on mites. In *Biological Control of Pests by Mites* (eds M.A. Hoy, G.L. Cunningham and L. Knutson), University of California Special Publication no. 3304, pp. 2–11.

Rosenthal, S.S. (1983) Current status and potential for biological control of field bindweed, *Convolvulus arvensis* with *Aceria convolvuli*. In *Biological Control of Pests by Mites* (eds M.A. Hoy, G.L. Cunningham and L. Knutson), University of California Special Publication no. 3304, pp. 57–60.

Royalty, R.N. and Perring, T.M. (1987) Comparative toxicity of acaricides to *Aculops lycopersici* and *Homeopronematus anconai* (Acari: Eriophyidae, Tydeidae). *J. Econ. Entomol.*, **80**, 348–51.

Ryba, J., Rodl, P., Bartos, L., Daniel, M. and Cerny, V. (1987) Some features of the ecology of fleas inhabiting the nests of the European suslik (*Citellus citellus* (L.). II. The influence of mesostigmatid mites of fleas. *Folia Parasitol.*, **34**, 61–8.

Sabelis, M.W. (1985) Predation on spider mites. In *Spider Mites, Their Biology, Natural Enemies and Control* (eds W. Helle and M.W. Sabelis), Elsevier, Amsterdam, vol. 1B, pp. 103–29.

Sabelis, W.M. and Dicke, M. (1985) Long-range dispersal and searching behaviour. In *Spider Mites: Their Biology, Natural Enemies and Control* (eds W. Helle and M.W. Sabelis), Elsevier, Amsterdam, vol. 1B, pp. 141–60.

Saito, Y. (1986) Prey kills predator: counter-attack success of a spider mite against its specific phytoseiid predator. *Exp. Appl. Acarol.*, **2**, 47–62.

Samarasinghe, S. and LeRoux, E.J. (1966) The biology and dynamics of the oystershell scale, *Lepidosaphes ulmi* (L.) (Homoptera: Coccidae), on apple in Canada. *Ann. Entomol. Soc. Quebec*, **11**, 206–92.

Santos, M.A. (1976) Evaluation of *Zetzellia mali* as a predator of *Panonychus ulmi* and *Aculus schlechtendali*. *Environ. Entomol.*, **5**, 187–91.

Santos, M.A. (1984) Effects of host plant on the predator–prey cycle of *Zetzellia mali* (Acari: Stigmaeidae) and its prey. *Environ. Entomol.*, **13**, 65–9.

Schroder, R.F.W. (1983) The potential use of mites in biological control of field crop pests. In *Biological Control of Pests by Mites* (eds M.A. Hoy, G.L. Cunningham and L. Knutson), University of California Special Publication no. 3304, pp. 36–40.

Sellers, W.F. and Robinson, G.G. (1950) The effect of the miticide Neotran® upon the laboratory production of *Aspidiotus lataniae* Signoret as a coccinellid food. *Can. Entomol.*, **82**, 170–3.

Severin, H.C. (1944) The grasshopper mite, *Eutrombidium trigonum* (Hermann), an important enemy of grasshoppers. *South Dakota Agric. Exp. Stn*, Tech. Bull. **3**, 1–36.

Sharma, R.D. (1971) Studies on the plant parasitic nematode *Tylenchorhynchus dubius*. *Meded. Landbouww. Wageningen*, **71**, 1–154.

Sinha, R.N. and Wallace, H.A.H. (1966) Association of granary mites and seed-borne fungi in stored grain and in outdoor and indoor habitats. *Ann. Entomol. Soc. Am.*, **59**, 1170–81.

Small, R.W. (1988) Invertebrate predators. In *Diseases of Nematodes* (eds G.O. Poinar Jr and H.-B. Jansson), CRC Press, Boca Raton, pp. 73–92.

Smiley, R.L. (1984) The ordinal and subordinal names of mites with a list of the mite pests of stored products. In *Proceedings of the Third International Working Conference on Stored-product Entomoloy*, Kansas State University, Manhattan, Kansas, pp. 37–43.

Smiley, R.L. and Knutson, L. (1983) Aspects of taxonomic research and services relative to mites as biological control agents. In *Biological Control of Pests by Mites* (eds M.A. Hoy, G.L. Cunningham and L. Knutson), University of California Special Publication no. 3304, pp. 148–64.

Smith, B.P. (1983) The potential of mites as biological control agents of mosquitoes. In *Biological Control of Pests by Mites* (eds M.A. Hoy, G.L. Cunningham and L. Knutson), University of California Special Publication no. 3304, pp. 79–85.

Smith, B.P. (1988) Host–parasite interaction and impact of larval water mites on insects. *Annu. Rev. Entomol.*, **33**, 487–507.

Smith, B.P. and McIver, S.B. (1984) The impact of *Arrenurus danbyensis* (Mullen) (Acari: Prostigmata; Arrenuridae) on a population of *Coquillettidia perturbans* (Walker) (Diptera: Culicidae). *Can. J. Zool.*, **62**, 1121–34.

Smith, I.M. and Oliver, D.R. (1986) Review of parasitic associations of water

REFERENCES

mites (Acari: Parasitengona: Hydrachnida) with insect hosts. *Can. Entomol.*, **118**, 407–72.

Sobhian, R. and Andres, L.A. (1978) The response of the skeletonweed gall midge, *Cystiphora schmidti* (Diptera: Cecidomyiidae), and gall mite, *Acaria chondrillae* (Eriophyidae) to North American strains of rush skeletonweed (*Chondrilla juncea*). *Environ. Entomol.*, **7**, 506–8.

Solomon, M.E. (1969) Experiments on predator–prey interactions of storage mites. *Acarologia*, **11**, 484–503.

Sorensen, J.T., Kinn, D.N., Doutt, R.L. and Cate, J.R. (1976) Biology of the mite *Anystis agilis* (Acari: Anystidae): a California vineyard predator. *Ann. Entomol. Soc. Am.*, **69**, 905–10.

Sorensen, J.T., Kinn, D.N. and Doutt, R.L. (1983) Biological observations on *Bdella longicornis*: A predatory mite in California vineyards (Acari: Bdellidae). *Entomography*, **2**, 297–305.

Stein, W. (1960) Untersuchungen uber die Möglichkeit einer Bekämpfung von Raubmilben in Zuchten Getreidenmotte (*Sitotroga cerealella* (Oliv.)) durch Anwendung von Akariziden. *Z. Pflanzenk. (Pflanzenp.) Pflanzens.*, **67**, 77–87.

Strickler, K. and Croft, B.A. (1985) Comparative rotenone toxicity in the predator, *Amblyseius fallacis* (Acari: Phytoseiidae), and the herbivore, *Tetranychus urticae* (Acari: Tetranychidae), grown on lima beans and cucumbers. *Environ. Entomol.*, **14**, 243–6.

Strong, R.G., Pieper, G.R. and Sbur, D.E. (1959) Control and prevention of mites in granary and rice weevil cultures. *J. Econ. Entomol.*, **52**, 443–6.

Sturhan, D. and Hampel, G. (1977) Pflanzenparasitische Nematoden als Beute der Wurzelmilbe *Rhizoglyphus echinopus* (Acarina, Tyroglyphidae). *Anz. Schädl. Pflanzensch. Umwelt.*, **50**, 115–18.

Sunderland, K.D. (1988) Quantitative methods for detecting invertebrate predation occurring in the field. *Ann. Appl. Biol.*, **112**, 201–24.

Swift, F.C. (1970) Predation of *Typhlodromus* (A.) *fallacis* on the European red mite as measured by the insecticidal check method. *J. Econ. Entomol.*, **63**, 1617–18.

Takafuji, A. and Chant, D.A. (1976) Comparative studies of two species of predaceous phytoseiid mites (Acarina: Phytoseiidae), with special reference to their responses to the density of their prey. *Res. Popul. Ecol.*, **17**, 255–310.

Tandon, P.L. and Lal, B. (1976) New record of predatory mites on mango mealy bug, *Drosicha mangiferae* Green (Margarodidae: Hemiptera). *Current Sci.*, **45**, 566–7.

Tanigoshi, L.K., Nishio-Wong, J.Y. and Fargerlund, J. (1984) *Euseius hibisci*: Its control of citrus thrips in southern California citrus orchards. In *Acarology VI* (eds D.A. Griffiths and C.E. Bowman), Ellis Horwood, Chichester, Vol. 2, pp. 717–24.

Tevis, L. and Newell, I.M. (1962) Studies on the biology and seasonal cycle of the giant velvet mite, *Dinothrombium pandorae* (Acari: Trombidiidae). *Ecology*, **43**, 497–505.

Thomas, W.P. and Chapman, L.M. (1978) Integrated control of apple pests in New Zealand. 15. Introduction of two predacious phytoseiid mites. *Proc. 31st NZ Weed & Pest Cont. Con.*, 236–43.

Thorvilson, H.G., Phillips, S.A. Jr, Sorensen, A.A. and Trostle, M.R. (1987) The straw itch mite, *Pyemotes tritici* (Acari: Pyemotidae), as a biological control agent of red imported fire ants, *Solenopsis invicta* (Hymenoptera: Formicidae). *Fla. Entomol.*, **70**, 439–44.

Tseng, Y.-H. (1979) Studies on the mites infesting stored food products on Taiwan. In *Recent Advances in Acarology* (ed. J.G. Rodriguez), Vol. 1, pp. 311–16.

Turnbull, A.L. and Chant, D.A. (1961) The practice and theory of biological control of insects in Canada. *Can. J. Zool.*, **39**, 697–753.

van de Bund, C.F. (1972) Some observations on predatory action of mites on nematodes. *Zesz. Probl. Postepow nauk Rolniczych*, **129**, 103–10.

van de Vrie, M., McMurtry, J.A. and Huffaker, C.B. (1972) Ecology of tetranychid mites and their natural enemies: a review. III. Biology, ecology, and pest status, and host-plant relations of tetranychids. *Hilgardia*, **41**, 343–432.

van Haren, R.J.F., Steenhuis, M.M., Sabelis, M.W. and de Ponti, O.M.B. (1987) Tomato stem trichomes and dispersal success of *Phytoseiulus persimilis* relative to its prey *Tetranychus urticae*. *Exp. Appl. Acarol.*, **3**, 115–21.

van Houten, Y.M., Overmeer, W.P.J., van Zon, A.Q. and Veerman, A. (1988) Thermoperiodic induction of diapause in the predacious mite, *Amblyseius potentillae*. *J. Insect Physiol.*, **34**, 285–90.

van Lenteren, J.C. and Woets, J. (1988) Biological and integrated pest control in greenhouses. *Annu. Rev. Entomol.*, **33**, 329–69.

van Zon, A.Q. and Wysoki, M. (1978) The effect of some fungicides on *Phytoseiulus persimilis* (Acarina: Phytoseiidae). *Entomophaga*, **23**, 371–8.

Wallace, M.M.H. (1954) The effect of DDT and BHC on the population of the lucerne flea, *Sminthurus viridis* (L.) (Collembola), and its control by predatory mites, *Biscirus* spp. (Bdellidae). *Aust. J. Agric. Res.*, **5**, 148–55.

Wallace, M.M.H. (1967) The ecology of *Smithurus viridis* (L.) (Collembola). I. Processes influencing numbers in pastures in Western Australia. *Aust. J. Zool.*, **15**, 1173–206.

Wallace, M.M.H. (1972) A portable power-operated apparatus for collecting epigaeic Collembola and Acari. *J. Aust. Entomol. Soc.*, **11**, 261–3.

Wallace, M.M.H. (1981) Tackling the lucerne flea and the red-legged mite. *J. Agric. West. Aust.*, **22**, 72–4.

Wallace, M.M.H. and Holm, E. (1983) Establishment and dispersal of the introduced predatory mite, *Macrocheles peregrinus* Krantz, in Australia. *J. Aust. Entomol. Soc.*, **22**, 345–8.

Wallace, M.M.H., Tyndale-Biscoe, M. and Holm, E. (1979) The influence of *Macrocheles glaber* on the breeding of the Australian bushfly, *Musca vetustissima* in cow dung. In *Recent Advances in Acarology* (ed. J.G. Rodriguez), Academic Press, New York, vol. 2, pp. 217–22.

Wallace, M.M.H. and Walters, M.C. (1974) The introduction of *Bdellodes lapidaria* (Acari: Bdellidae) from Australia into South Africa for the biological control of *Sminthurus viridis*. *Aust. J. Zool.*, **22**, 505–17.

Walter, D.E. (1988) Predation and mycophagy by endeostigmatid mites (Acariformes: Prostigmata). *Exp. Appl. Acarol.*, **4**, 159–66.

Walter, D.E., Hunt, H.W. and Elliott, E.T. (1988) Guilds or functional groups?

REFERENCES

An analysis of predatory arthropods from a shortgrass steppe soil. *Pedobiologia*, **31**, 247–60.

Wassenaar, D.P.J. (1988) Effectiveness of vacuum cleaning and wet cleaning in reducing house-dust mites, fungi and mite allergen in a cotton carpet: a case study. *Exp. Appl. Acarol.*, **4**, 53–62.

Weires, R.W. and Smith, G.L. (1978) Apple mite control, Hudson Valley, 1977. *Insectic. Acaric. Tests*, **3**, 42–3.

Weiser, J. (1963) Über Massenzuchten von *Pyemotes*-Milben. *Beit. z. Entomol.*, **13**, 547–51.

Welbourn, W.C. (1983) Potential use of trombidioid and erythraeoid mites as biological control agents of insect pests. In *Biological Control of Pests by Mites* (eds M.A. Hoy, G.L. Cunningham and L. Knutson), University of California Special Publication no. 3304, pp. 103–40.

Wharton, G.W. and Arlian, G. (1972) Predatory behaviour of the mite *Cheyletus aversor*. *Anim. Behav.*, **20**, 719–23.

White, N.D.G. and Laing, J.E. (1977a) Field observations of *Zetzellia mali* (Ewing) (Acarina: Stigmaeidae) in southern Ontario apple orchards. *Proc. Entomol. Soc. Ontario*, **108**, 23–30.

White, N.D. and Laing, J.E. (1977b) Some aspects of the biology and a laboratory life table of the acarine predator *Zetzellia mali*. *Can. Entomol.*, **109**, 1275–81.

Willis, R.R. and Axtell, R.C. (1968) Mite predators of the house fly: a comparison of *Fuscuropoda vegetans* and *Macrocheles muscaedomesticae*. *J. Econ. Entomol.*, **61**, 1669–74.

Woodroffe, G.E. (1953) An ecological study of the insects and mites in the nests of certain birds in Britain. *Bull. Entomol. Res.*, **44**, 739–72.

Woolhouse, M.E.J. and Harmsen, R. (1984) The mite complex of the foliage of a pesticide-free apple orchard: population dynamics and habitat associations. *Proc. Entomol. Soc. Ont.*, **115**, 1–11.

Young, O.P. and Welbourn, W.C. (1987) Biology of *Lasioerythraeus johnstoni* (Acari: Erythraeidae), ectoparasitic and predaceous on the tarnished plant bug, *Lygus lineolaris* (Hemiptera: Miridae), and other arthropods. *Ann. Entomol. Soc. Am.*, **80**, 243–50.

Zakhavtkin, A.A. (1959) Fauna of the USSR, Arachnoidea, Vol. VI (1), Tyroglyphoidea (Acari). American Institute of Biological Sciences, pp. 1–573.

Žďárková, E. (1986) Mass rearing of the predator *Cheyletus eruditus* (Schrank) (Acarina: Cheyletidae) for biological control of acarid mites infesting stored product. *Crop. Prot.*, **5**, 122–4.

Žďárková, E. and Horak, E. (1987) Contact acaricides may not restrain effectiveness of the biological control against stored food mites. *Acta. Entomol. Bohemosl.*, **84**, 414–21.

Žďárková, E. and Pulpan, J. (1973) Low temperature storage of the predatory mite *Cheyletus eruditus* (Schrank) for future use in biological control. *J. Stored Prod. Res.*, **9**, 217–20.

43
Recommendations for further reading

By way of summation, we close this book with some recommendations for further study. The following recommendations were made by contributors to the conference on pest control by mites (Hoy *et al.*, 1983): that further exploration for predaceous mites be undertaken, as well as basic research on mite systematics and biology, development of mass-rearing methods and quality control, on the conduct of field experiments with promising candidates while employing appropriate controls, and finally devising better (and more uniform) evaluation procedures and means of integrating the emerging, promising acarine biocontrol agents into existing pest management programmes. In addition to these recommendations, we would like to make several of our own, besides those made in the preceding text.

1. The interactions between predaceous mites and their pest prey should be studied on the wild plants on which they occur in nature. Neither of these groups of arthropods has evolved on agricultural plants and their present-day interactions probably reflect patterns of behaviour and dispersal better suited to other plants (as well as to smaller pest populations). Data on such interactions could assist in predicting the performance of exotic acarine biocontrol agents and help plant breeders in selecting for traits which would enhance the ability of commercial plants to obtain their 'bodyguards' (Dicke and Sabelis, 1988). Schicha (1987) noted that none of the phytoseiids introduced into Australia have been found on native plants, although these predators had become widely distributed there. This point needs confirmation from other parts of the world, because of its obvious implications for spider mite control on indigenous crops.
2. Predaceous mites are sometimes rendered ineffective by extreme climatic conditions, especially low humidities and high temperatures. Suitable genetic improvement projects (as reviewed and advocated for phytoseiids by Hoy (1985), and for Macrochelidae by Anderson (1983) should therefore be undertaken. The search for more wild stock of 'domesticated' predators, and its incorporation into cultures which have been mass-reared for many generations is especially recommended.

RECOMMENDATIONS FOR FURTHER READING

3. The ability of some host/prey species to protect themselves against acarine biocontrol agents should be further investigated. Such protection could be due to larger prey size, as noted above, or to specific interactions. Lanciani (1988) described how mosquito larvae dislodged the parasitic larvae of *Arrenurus pseudotenuicollis* Wilson and even consumed them. Nymphs of *Thrips palmi* Karny frequently managed to evade attacks of *Amblyseius okinawanus* Ehara by rapid up-and-down body movements, tactics which were unsuccessful against *A. barkeri* (Kajita, 1986). It was not clear whether this was due to specific predator size or other factors.
4. The effect of the interspecific relationships between predaceous mites on their ability to control pests should be studied, as well as their potential ability to interfere with or even hinder other biological control agents. The native (to the western US) phytoseiid *Typhlodromus occidentalis* fed on eggs and larvae of the introduced *P. persimilis*, while the latter showed no reciprocal behaviour. *T. occidentalis* was therefore considered to be a better candidate for the control of *T. urticae* on hops (Pruszynski and Cone, 1972). Phytoseiids were unable to control spider mites on tea when stigmaeids occurred with them at comparable densities (Lo, 1986). Burnett (1977) reported that when *Cheyletus eruditus* and *Blattisocius dentriticus* (Berlese) co-habited in the same cultures, the former eliminated the latter. The interference of *Pyemotes* in the biological control of a coconut pest in Fiji has often been cited (Cock, 1986). This mite was also regarded by Traboulsi (1968) as a 'threat' to *Aphytis*, an important natural enemy of armoured scale insects. Predaceous mites may well 'control' spider mites or eriophyids introduced for the purposes of weed control. D.E. Walter (personal communication) stated that many mesostigmatid mites which feed on nematodes could also affect entomogenous nematodes introduced for pest control. This suggests that predaceous mites may be a group of biocontrol agents about which more will need to be known before they are freely introduced into new habitats.
5. Appropriate predaceous mites should be studied in regard to their ability to affect pests in unusual ways, including spread of pest diseases (Newell and Ryckman, 1966; Schabel, 1982), inhibiting host copulation (Moya Borja, 1981) or by competition and interference (Croft and Hoying, 1977; Ryba *et al.*, 1987).
6. Efforts should be made to try more species of mites for weed and nematode control and even for vectoring weed and nematode diseases.
7. Exploration for acarine predators should be undertaken in the tropics, especially in and near areas of emerging agriculture.

8. Several mite species, such as the mesostigmatids *Varroa jacobsoni* Oudemans and *Tropilaelaps clareae* Delfinado and Baker, have recently become major pests of honey bees, capable of totally destroying bee colonies (Needham *et al.*, 1988). Studies of interrelationships between these mites and their hosts may well supply basic insights into how mites control insect populations.

9. A documentation centre for acarine biocontrol agents should be established to document all efforts made in this area, even unsuccessful ones. Information on the systematics and biology of predaceous mites and their reactions to pesticides should also be stored there, listing both target pests and acarine agents. Relevant data are often 'submerged' or even lost due to ignorance about systematics and the specific interests of acarologists.

We wrote in the Preface that it was our hope that this book will promote further research about the potential of mites in the biological control of pests. We conclude by adding that we also hope for more theoretical studies of these interrelationships, along the lines noted above, which will, in turn, lead to additional advances in the biological control of pests by mites.

REFERENCES

Anderson, J.R. (1983) Mites as biological control agents of dung-breeding pests: practical considerations and selection for pesticide resistance. In *Biological Control of Pests by Mites* (eds M.A. Hoy, G.L. Cunningham and L. Knutson) University of California Special Publication no. 3304, pp. 99–102.

Burnett, T. (1977) Biological models for two acarine predators of the grain mite, *Acarus siro* L. *Can. J. Zool.*, **55**, 1312–23.

Cock, M.J.W. (1986) Requirements for biological control: an ecological perspective. *Biocon. News Inform.*, **7**, 7–16.

Croft, B.A. and Hoying, S.A. (1977) Competitive displacement of *Panonychus ulmi* (Acarina: Tetranychidae) by *Aculus schlechtendali* (Acarina: Eriophyidae) in apple orchards. *Can. Entomol.*, **109**, 1025–34.

Dicke, M. and Sabelis, M.W. (1988) How plants obtain predatory mites as bodyguards. *Neth. J. Zool.*, **38**, 148–65.

Hoy, M.A. (1985) Recent advances in genetics and genetic improvement of the Phytoseiidae. *Annu. Rev. Entomol.*, **30**, 345–70.

Hoy, M.A., Cunningham, G.L. and Knutson, L. (1983) *Biological Control of Pests by Mites*. University of California Special Publication no. 3304, p. 185.

Kajita, H. (1986) Predation by *Amblyseius* spp. (Acarina: Phytoseiidae) and *Orius* sp. (Hemiptera: Anthocoridae) on *Thrips palmi* Karni (Thysanoptera: Thripidae). *Appl. Entomol. Zool.*, **21**, 482–4.

RECOMMENDATIONS FOR FURTHER READING

Lanciani, C.A. (1988) Defensive consumption of parasitic mites by *Anopheles crucians* larvae. *J. Am. Mosquito Control Assoc.*, **4**, 195.

Lo, P.K.-C. (1986) Present status of biological control of mite pests in Taiwan. *Plant Prot. Bull. (Taiwan)*, **28**, 31–9.

Moya Borja, G.E. (1981) Effects of *Macrocheles muscaedomesticae* (Scopoli) on the sexual behaviour and longevity of *Dermatobia hominis* (L. Jr.). *Rev. Brasil Biol.*, **41**, 237–41.

Needham, G.R., Page, R.E. Jr, Delfinado-Baker, M. and Bowman, C.E. (eds) (1988) *Africanized Honey Bees and Bee Mites*. Ellis Horwood, Chichester, p. 572.

Newell, I.M. and Ryckman, R.E. (1966) Species of *Pimeliaphilus* (Acari: Pterygosomatidae) attacking insects, with particular reference to the species parasitizing *Triatominae* (Hemiptera: Reduviidae). *Hilgardia*, **37**, 403–36.

Pruszynski, S. and Cone, W.W. (1972) Relationships between *Phytoseiulus persimilis* and other enemies of the twospotted spider mite on hops. *Environ. Entomol.*, **1**, 431–3.

Ryba, J., Rodl, P., Bartos, L., Daniel, M. and Cerny, V. (1987) Some features of the ecology of fleas inhabiting the nests of the European suslik (*Citellus citellus* (L.). II. The influence of mesostigmatid mites of fleas. *Folia Parasitol.*, **34**, 61–8.

Schabel, H.G. (1982) Phoretic mites as carriers of entomopathogenic fungi. *J. Invert. Pathol.*, **39**, 410–12.

Schicha, E. (1987) *Phytoseiidae of Australia and Neighbouring Areas*. Indira Publishing House, p. 187.

Traboulsi, R. (1968) Predateurs et parasites d'*Aphytis* (Hym., Aphelinidae). *Entomophaga*, **13**, 345–55.

Index

(References to Figures in italics)

Acaridae *19, 21, 23,* 46–7, 130
Acariformes 15, 17
Acaronemus destructor 98
Acaropsis 147
Acarus 58
Aceria 62, 132
    Aceria chondrillae 62, 132, *136*
    Aceria convolvuli 62, *131*
Acremonium zonatum 68
Acrididae 88, 102, *112–13*
Acroptilon repens 132
Aculus lycopersici 8, 96, 104
Aculus schlechtendali 63, 96, 138
Adactylidium 94
Adaptability of predators 141
Aedes 51
    Aedes ventrovittis 72
Aerial release 131
Africa 69, 70, 72, 100, 113, 131
Agistemus 96
    Agistemus exsertus 96, *130*
    Agistemus longisetus 96
Allothrombium monspessulanum 102, *138*
Allothrombium pulvinum 102
Alternative prey or hosts 53, 104, 126, 129–30
see also Supplementary diet
Amblyseius 83, 130
    Amblyseius barkeri (=*mckenziei*) 84, 146, 163
    Amblyseius cucumeris 84
    Amblyseius fallacis 127, 139
    Amblyseius okinawaensis 163
    Amblyseius potentillae 119, 120–1
    Amblyseius swirskii 114

Amblyseius teke 130
Amblyseius victoriensis 84
Ambrosia 62
Anal plate 18, *20, 22–3, 24,* 40, *43*
Androlaelaps 74
    Androlaelaps casalis 40
    Androlaelaps fahrenholzi 75
Anopheles crucians 50, 136, *138–9*
Anopheles farauti 76
Antarctic microarthropods 137
Anthonomus grandis 5, 129
Ants 93, 134, 135
Anystidae 19, 32, 48–9, 124, 140
Anystis 19, 32, 49, 120, 142, 145
    Anystis agilis 48, *143–4*
    Anystis baccarum 48, 119, 129
    Anystis salicinus 49
Aonidiella aurantii 56
Aphelenchus avenae 53
    see also Nematodes
Aphids 48, 54, 102, 138
Aphytis 125, *163*
Apple 5, 61, 63, 65, 70, 96, 119–20, 123, 128, 138, 140
Arctoseius cetratus 52
Argentina 68, 114
Armoured scale insects *see* Diaspididae
Arrenuridae *35–8*, 50–1
    Arrenurus 37, 38, 50, *51*
    Arrenurus danbyensis 50, 51, 116
    Arrenurus liberiensis 35, 36, 37
    Arrenurus madaraszi 50
    Arrenurus pseudoaffinis 37
    Arrenurus pseudotenuicollis 163

INDEX

Artificial media for predators 94, 130
Ascidae *16, 41–2, 44, 52–3,* 118, 124
Astigmata 14, 17, 18, 20, 23, 58, 111, 137
Australia 6, 54, 62, 79, 84, 100, 131, 135, 136, 162
Azadirachta indica 128

Balaustium 34, 124
  Balaustium murorum 65
  Balaustium putmani 64
Bark beetles *see* Scolytidae
Bdella depressa 54
Bdella distincta 55
Bdella lignicola 118
Bdella longicornis 54
Bdellidae 19, 27, 54–5, 118, 124, 131, 137–8
Bdellodes 143
Bdellodes lapidaria 6, 54, 129, 134, 142
Bean 121, 122
Bechsteinia 32
Behavioural resistance 128
Bermuda 70, 131, 145
Biological control 1–3, 7, 6, 54, 58, 62, 78, 96, 117, 120, 133–41
  augmentation 2, 58, 133–4
  conservation 2, 48
  introduction 2, 5, 49, 54, 62, 69, 70, 78–9, 100, 102, 117, 133–4, 162
Biting midges *see* Ceratopogonidae
Black flies *see* Simuliidae
Blattisocius 42, 144
  Blattisocius dentriticus 163
  Blattisocius keegani 16, 41, 42, 44
  Blattisocius tarsalis 52, 124, 135, 142, 143, 145
Bochartia 114
Bourletiella arvalis 6, 54
Brevipalpus phoenicis 96, 142
Bryobia praetiosa 26, 31, 54

Cacti, 6, 100
Cages and barriers 134
Calepitrimerus vitis 61
Camerobiidae 33, 34, 56–7, 114

Canada 5, 50, 70, 134, 138
Carotene 120–1
Carpoglyphidae 19, 21
Carpoglyphus lactis 19, 21
Caudacheles lieni 118
Cedar 145
Centaurea repens 62
Ceratopogonidae 72, 76, 115
Cheiroseius 52
Cheletogenes ornatus 59
Cheletomimus binus 118
Cheletomimus bisetosa 118
Cheletomorpha lepidopterorum 31, 118
Cheletonella pilosa 118
Chelicerae 17, *18, 19, 24, 27, 30, 32, 40,* 62, 66, 86, 88, 91, 93, 96, 100, 102, 104
Cheyletidae 18, 26, 30, *31,* 58–60, 114, 118, 124, 126, 140
Cheyletus 6, 136, 143
  Cheyletus eruditus 6, 18, 26, 30, 58, 118, 130, 142, 144, 146, 163
  Cheyletus fortis 118
  Cheyletus malaccensis 118
  Cheyletus trouessarti 118
Chilocorus 70, 131, 143, 145
  Chilocorus bipustulatus 130
China 102
Chironomidae 86
Chrondrilla juncea 62, 132
Chortoicetes terminifera 88, 112
Chrysomelobia labidomerae 88, 132
Chrysomphalus aonidum 66
Citellophilus simplex 75
Citellus citellus 75
Citrus 48, 59, 66, 70, 84, 104, 114, 128
Coccipolipus epilachnae 25, 89
Coccidae 114
Coccidea 104, 114
Coccinellidae 70, 89, 143
Cockroaches 91, 146
Coconut 74, 163
Cold storage 58, 94, 129
Collection of predaceous mites 129
Collembola 6–7, 49, 54, 81, 131, 135, 137–8, 142–3
Competition with pests 2, 163

INDEX

Convolvulus arvensis 62, 131
Coquillettidia 51
   Coquillettidia perturbans 50, 138
Corn rootworms 46, 74, 138
Cotton 5, 93–4
Coxal plates *38*, 50, 72, 76
Crispa metopica 33–4
Cryptostigmata 17, 18, 20, 22, 111, 137
Cucumber 122
Culex 51
   Culex pullus 76
Culicidae 6, 8, 50–1, 52, 72, 76, 86, 115–16, 136, 138–9, 163
Cunaxa 27
   Cunaxa capreolus 61
   Cunaxa taurus 118
   Cunaxa womersleyi 118
Cunaxidae 27, 28, 61, 118
Cunaxoides oliveri 61
Cunaxoides parvus 61
Curculionidae 68, 129
Czechoslovakia 58

Daktulosphaira vitifolii 5, 46
Density dependence 53, 141, 146–7
Dermatitis 93
Dermatophagoides 8
   Dermatophagoides farinae 21
   Dermatophagoides pteronyssinus 17, 21, 58
Deuterothyas variabilis 35
Diabrotica longicornis 74, 138
Diabrotica virgifera 74, 138
Diabrotica undecimpunctata howardi 46
Diaspididae 7, 55, 56, 59, 66, 70–1, 114, 130, 143, 145–6
Dinothrombium pandorae 112
Diptera 8, 115, 142
Dispersal of predators 52, 78, 83, 142–3, 145, 163
Displacement of residential predators 127–8
Ditylenchus 46
   see also Nematodes
Documentation 164
Domatia *see* Leaf domatia
Drosicha mangiferae 114

Dung *see* Manure
Dung beetles *see* Scarabaeidae

Egypt 93, 94
Eichhornia crassipes 68, 100
El Salvador 89
Empodium *17, 18, 20, 23, 28, 29, 31,* 43, 66, 93
Ephestia cautella 52, 124, 135, 142
Epigynial plate *20, 23, 24, 37, 42, 43,* 50, 52, 68, 72, 74, 81
Epilachna varivestis 8, 89
Eriophyes boycei 62
Eriophyes ficus 25
Eriophyidae 25, 62–3, 96, 104, 131–2, 163
Erythraeidae 34, 64–5, 124, 140
Erythroneura elegantula 48
Eucheyletia reticulata 118
Eulaelaps stabularis 75
Eupalopsellidae *33*, 66–7, 114
Eupalopsis maseriensis 33
Euseius 144
   Euseius hibisci 84, 142
Eustigmaeus (=*Ledermuelleria*) 96
Eutetranychus orientalis 61
Eutrombidium locustarum 102, 112
Evaluation of predator efficacy 133–40, 162
Exploration for predators 162
Eyes 27, 35, 72, 76, 100

Fannia canicularis 78, 106
Fiji 69, 163
Filth flies 7, 136, 143
Fleas *see* Siphonaptera
Fovea pedalis *see* Leg grooves
France 5, 138
Functional response 139–40, 141
Fungal diseases 68, 120
Fuscuropoda 41
   Fuscuropoda agitans 16
   Fuscuropoda vegetans 106, 130, 146

Gall mites *see* Eriophyidae
Galls 62, 136
Galumna virginiensis 22

INDEX

Galumnidae 15, 22, 68–9
Gamasellus racovitzai 137
Geckobiella texana 19, 32
Genetic improvement 7, 83, 128, 162
Genital acetabula 37, 50, 76
  see also Epigynial plate
Genital plate see Epigynial plate
Germany 61
Glycyphagidae 19, 21
Glyptholaspis confusa 134
Gnathosoma 27, 56
  see also Palpus, Thumb-claw process
Gohieria fusca 21
Grallacheles bakeri 118
Grape pests 5, 46, 48, 54, 61
Grasshoppers see Acrididae
  see also Locusts
Greece 62, 136
Greenhouse crops 83–4, 144, 146
Gut contents of predators
  chromatography 136–7
  electrophoresis 137
  pigments 137
  visual examination 137
Gynaikothrips ficorum 94

Haematobia irritans exiqua 79
Haemogamasus nidi 75
Haemogamasus pontiger 74
Haller's organ 39, 40
Halotydeus destructor 49
Hemicheyletia arecana 118
Hemicheyletia bakeri 58, 59, 139
Hemisarcoptes 7, 24, 70, 71, 114, 131, 143, 144, 145
  Hemisarcoptes coccophagus 24, 70, 130, 144
  Hemisarcoptes malus 5, 15, 18, 24, 70, 71, 123–5, 131, 134, 138, 142
Hemisarcoptidae 15, 18, 24, 70–1, 114, 123, 124
Heterodera 46
  see also Nematodes
Honey bee mites 164
Honeydew 104, 114, 120, 123
Hop 146
Horse flies see Tabanidae

Hydryphantidae 35, 38, 72–3
Hypoaspis 74
  Hypoaspis aculeifer 74, 111
Hypopodes 14, 18, 19, 70, 130–1, 143
Hysterosoma 29

Ilex europaeus 100
Identification and systematics of predators 8, 132, 162, 164
India 66, 69, 114
Indonesia 96
Insect growth regulators (IGRs) 128
  see also Pesticides
Insect–mite associations 2
Instantaneous death rate 72, 139
Integument 20, 36, 72, 76, 86, 98
Interference with pests 163
Intraspecific predator variability 62, 132
Intrinsic rate of increase 96, 114
  see also Power of increase
Iponemus 98
Ips 98
Iridomyrmex humilis 135
  see also Ants
Israel 59, 66, 70
Italy 62
Ixodida 17, 39, 40, 49

Kairomones 68, 83, 121, 143
Kenya 52, 88, 112, 135

Labidomera clivicollis 88
Laelapidae 40, 43, 74–5, 136, 138
Laelaps 43
Lasioerythraeus johnstoni 64, 146
Lasioseius allii 118
Lasioseius martini 118
Lasioseius paraberlesi 52
Lasioseius pencilliger 111
Lasioseius scapulatus 52, 53, 134
Lasioseius sugarawai 118
Leaf domatia 119, 121
Leaf-surface texture 119
  see also Trichomes
Leg grooves 16, 41, 106

INDEX

Lepidoglyphus destructor 19
Lepidoptera 52, 65, 135, 142–43
Lepidosaphes ulmi 5, 61, 70, 123
Leptinotarsa decemlineata 8, 88
Leptus 114
Life tables 138
Limnesiidae 35, 36, 38, 76–7
Limnesia 36, 38
   Limnesia jamurensis 76
   Limnesia lembangensis 36
   Limnesia lucifera uniseta 35
   Limnesia pinguipalpis 36
Locusta migratoria 88, 112
Locustacarus 88, 112
Locusts 88, 112–13
   see also Grasshoppers
Longidorus 46
   see also Nematodes
Lorryia formosa 104
Lycoriella auripila 52
Lygus lineolaris 64, 146

Macrocheles 78, 79, 142, 143, 144, 145
   Macrocheles glaber 145, 146
   Macrocheles muscaedomesticae 42, 43, 78, 138
   Macrocheles peregrinus 78, 131
Macrochelidae 42, 43, 78–80, 125, 130, 135–6, 162
Mango 114
Manure
   substrate for filth flies 78, 81, 106, 134, 136, 142, 144–6
   promoter of predator activity 74, 138
Margarodidae 114
Mass-rearing 5, 71, 129–30
Melanaspis glomerata 66
Melichares agilis 118
Melichares mali 118
Mesostigmata 2, 17, 40, 41, 111, 118, 137, 163
Mexico 5, 88, 129, 131
Microtrombidium 34
Mideopsidae 37–8
Mideopsis 37, 38
   Mideopsis fibrosa 37
Mineral fertilizers 120

Models of predator–prey interactions 140
Mononychellus tanajoa 7, 130
Mosquitoes see Culicidae
Moths see Lepidoptera
Musca autumnalis 78
Musca domestica 78, 81, 106
Musca vetustissima 78, 146
Mushrooms 52

Nectar 120
Nematodes 2, 46–7, 53, 69, 74, 104, 106, 111, 130, 133–4, 143, 163
Neochetina eichhorniae 68
Neomolgus capillatus 54
Neophyllobius 56, 133
   Neophyllobius ambulans 56
   Neophyllobius lorioi 33
Neopsylla setosa 75
Nests 74–5, 136
Netherlands 84
New Zealand 56, 88, 96, 100, 112, 128, 131, 133
Norway 75
Numerical predator/pest ratios 49, 58, 102, 137–8
Numerical response 52, 59, 139–40, 141

Observations on predator activity 136–7, 139
Opuntia 100, 136
Opuntia inermis 6, 100
Oribatulidae 17, 20
Orthogalumna terebrantis 15, 22, 68
Oryctes rhinoceros 74
Outbreaks of pests 123, 127

Pakistan 51
Palm 70
Palpus 18, 26–7, 32–3, 35–7, 39, 40, 50, 54, 56, 58, 61, 66, 72, 86
   see also Gnathosoma, Thumb-claw process
Panisopsis 72
Panonychus citri 48

INDEX

Panonychus ulmi 63, 64, 120, 121, 138, 147
Papua New Guinea 76
Parabonzia bdelliformis 27, 28
Parasitidae 42, 43, 81–2
Parasitiformes 15, 17, 39
Parlatoria blanchardi 70
Pasture 6, 54, 135, 142, 144
Peach 122
Pecan 128
Pectiniphora gossypiella 93
Pergalumna 69
Pergamasus 81
   Pergamasus longicornis 81
   Pergamasus quisquiliarum 81
Peritremes 16, 30, 40–2, 52, 56, 78, 91
Pesticides 63, 64, 70, 122–8, 135, 144, 164
   acaricides 58, 104, 122, 123–5
   fungicides 125–6, 128
   herbicides 126
   insecticides 58, 123–8
Pesticide-resistant predators 79, 83, 125, 127–8, 132, 143, 144
   see also Variable effects of pesticides
Pesticide detoxification enzymes 122
Pest diseases 2, 79, 91, 163
Pests 1, 2
   see also Nematodes, Weeds
Pests of laboratory cultures 66, 71, 91, 124–5, 147
Phoresy 2, 7, 78, 106, 145
Phytonomus pallidus 84
Phytoseiidae 1, 6, 7, 44, 54, 83–5, 114, 119, 122, 123–8, 130, 140, 141–6, 162
Phytoseiulus persimilis 83, 120–1, 127, 133, 142, 144–6, 163
Phytoseius ferox 44
Pimeliaphilus 91
   Pimeliaphilus cunliffei 30, 146
   Pimeliaphilus plumifer 91
Pine 48, 98
Piona 36, 38, 86
   Piona catatama 37
   Piona nodata 86, 115
Pionidae 36, 37, 38, 86–7

Plant diseases 63
Plant resistance to pests 120–1
Platyseius 52
Podapolipidae 25, 88–90, 112
Podapolipoides grassi 88, 112
Poecilochirus monospinosus 81
Pollen 74, 104, 120, 126, 143
Pot experiments 2, 52, 74, 111
Power of increase 141, 144–5
   see also Intrinsic rate of increase
Prey
   enrichment 136
   paralysis 56
Proctolaelaps pygmaeus 118
Propodosoma 20, 27–8, 30, 33–4, 91
Prostigmata 2, 17, 18, 25, 111, 118, 126, 137, 142
Pseudococcidae 114
Pseudostigmatic organ *see* Sensillae
Psylla pyricola 64
Pteromorphs 22, 68
Pterygosomatidae 19, 30, 32, 91–2, 125
Pyemotes (=*Pediculoides*) 5, 93, 94, 114, 129–31, 163
   Pyemotes herfsi 93
   Pyemotes tritici 15, 19, 29, 93, 94, 134
Pyemotidae 15, 29, 93–5, 146
Pyroglyphidae 17, 21

Quadraspidiotus ostreaeformis 56
Quality control of predators 130, 162

Races of predaceous mites 62, 88
Removal of predators 134–6
Rhizoglyphus echinopus 19, 23, 46, 47, 133
Rhizoglyphus robini 23
Rhodacaridae 2
Rhodacarus roseus 111
Rice 52
Russia 49, 62

Saissetia oleae 104, 114
Saniosulus nudus 66, 147
Scale insects *see* Coccoidea
Scarabaeidae 74, 78–9, 106

INDEX

Scheloribates 17, 20
Schizolachnus piri-radiata 48
Schizotetranychus celerius 146
Scirtothrips citri 48, 84, 127
Scolytidae 6, 7, 98
Scutigerella immaculata 81
Searching ability and strategy 139, 141–4, 147
Secondary chemicals 121, 127
Sensillae *15–16, 20, 27–9*, 33, *34*, 54, 61, 64, 93, 98, 102, 104
Sex ratio 94, 104
Shipment of natural enemies 5, 131
Simuliidae 2, 72, 115
Siphonaptera 75, 136
Size differences 64, 115, 144–6, 163
Sminthurus viridis 6, 49, 54, 135
Soil arthropods 2, 8, 46, 53, 74, 81, 106, 111, 112
Solenopsis invicta 93, 134
Sooty mould 104, 114
South Africa 6, 54, 56, 79, 131
Spatial coincidence and heterogeneity 50, 54, 74, 81, 138–40, 142
Specificity of predators 53, 131, 141–6
Sperchon 2, 115, 116
Sperchontidae 2
Spider mites *see* Tetranychidae
Spinibdella bifurcata 19, 27
Springtails *see* Collembola
Steneotarsonemus spinki 52
Stigma *15–16*, 17, *40–1*, 52, 74, 78, 81, 83, 106
Stigmaeidae *34*, 39, 96–7, 120, 123, 125–6, 140
Stigmaeus glypticus 34
Stomoxys calcitrans 78
Stored product pests 2, 6, 46, 52, 58, 81, 93, 117–18, 124, 135, 142–4, 146
Stratiolaelaps 74
Strawberry 59, 84, 120
Strains *see* Races of predaceous mites
Stylostome 50, 139
Sugar cane 66
Suidasia pontifica 19
Supplementary diet 63, 98, 114, 120
  see also Alternative prey

Sweden 111
Systematics *see* Identification and systematics of predators
Synergists 128

Tabanidae 39, 72, 115
Taiwan 52, 117–18
Tarsal claw *18–19*, 29
  see also Thumb claw process
Tarsonemidae 29, 98–9
Tarsonemus myceliophagus 52
Tarsonemus scaurus 29
Tea 96, 163
Tenuipalpidae 96, 98
Tetranychidae 7, 26, 30, *31*, 48, 54, 58, 61, 83, 96, 98, 100–1, 104, 120–2, 123, 133–40, 141–6, 162
Tetranychus 100
  Tetranychus desertorum (=*opuntiae*) 6, 100, 136
  Tetranychus lintearius 100
  Tetranychus urticae 48, 83, 120, 121, 127, 146, 163
Thyas 72
  Thyas barbigera 72, 115
  Thyas stolli 35
Thrips *see* Thysanoptera
Thrips palmi 163
Thrips tabaci 84, 130, 146
Thumb-claw process 26, 48, 54, 58, 64, 66, 91, 96, 100
Thysanoptera 14, 146
Ticks *see* Ixodida
Tokelau 74
Tomato 104, 121, 125
Toxins 91, 93, 146
Triatoma 91
Triatomidae 91
Tribolium confusum 93
Trichobothrium *see* Sensillae
Trichomes 121
Trombidiidae 16, *34*, 102–3, 112–13
Tropilaelaps clareae 164
Tydeidae 18, 26, *28*, 29, 104–5, 114, 125, 140
Tydeus 18, 26, 28
Tydeus starri 28

INDEX

Tylenchorhynchus dubius 74
    see also Nematodes
Typhlodromus caudiglans 127
Typhlodromus occidentalis 83, 120, 128, 146, 163
Typhlodromus pyri 83, 127, 137, 138, 142, 147
'*Tyroglyphus phylloxerae*' 5, 131
Tyrophagus 46, 81
Tyrophagus putrescentiae 21, 23, 46, 47
Tyrrellia circularis 76

Uropodidae *16*, *41*, 106–7
USA 5, 62, 88–9, 112, 131

Variable effects of pesticides 126–7
Varroa jacobsoni 163
Venturia inequalis 125

Vicia faba 121
Vineyards *see* Grapes
Vitamin A *see* Carotene
Voracity of predators 142–4
Voucher specimens 132

Water mites, 6, 8, 38
    see also Arrenuridae, Hydryphantidae, Limnesiidae and Pionidae
Webbing 100, 120, 142–3, 146
Weeds 1, 2, 6, 62–3, 68, 100, 124, 126, 131, 136, 147, 163
Weevils *see* Curculionidae

Zeiraphera diniana 65
Zetzellia 96
    Zetzellia mali 96, 120, 138